J. A. Lintner

Secont Report on the Injurious and other Insects of the State of

New York

J. A. Lintner

Secont Report on the Injurious and other Insects of the State of New York

ISBN/EAN: 9783741123511

Manufactured in Europe, USA, Canada, Australia, Japa

Cover: Foto ©Klaus-Uwe Gerhardt /pixelio.de

Manufactured and distributed by brebook publishing software
(www.brebook.com)

J. A. Lintner

Secont Report on the Injurious and other Insects of the State of New York

SECOND REPORT

ON THE

INJURIOUS AND OTHER INSECTS

OF THE

STATE OF NEW YORK.

Made to the Legislature, Pursuant to Chapter 377 of the Laws of 1881.

By J. A. LINTNER,

State Entomologist.

ALBANY:

WEED, PARSONS AND COMPANY,

LEGISLATIVE PRINTERS.

1885.

No. 162.

IN ASSEMBLY,

APRIL 2, 1885.

SECOND REPORT

OF THE STATE ENTOMOLOGIST, ON THE INJU-
RIOUS AND OTHER INSECTS OF THE STATE OF
NEW YORK.

OFFICE OF THE STATE ENTOMOLOGIST, }
ALBANY, *March* 12, 1885. }

To the Hon. GEORGE Z. ERWIN,

Speaker of the Assembly :

SIR — I have the honor to transmit herewith, to the Legisla-
ture, pursuant to Chapter 377 of the Laws of 1881, the Second
Report on the Injurious and Other Insects of the State of New
York.

I remain very respectfully, your obedient servant,

J. A. LINTNER,
State Entomologist.

TABLE OF CONTENTS.

INJURIOUS LEPIDOPTEROUS INSECTS.

INJURIOUS HEMIPTEROUS INSECTS.

REPORT.

INTRODUCTION.

To the Legislature of the State of New York :

The Entomologist, in presenting his Second Report on the Injurious and other Insects of the State of New York, begs leave to state :

The delay of nearly a year in the printing of the First Report afforded the opportunity of earlier publication of considerable material that had been prepared for the following report. Much of this material was accordingly incorporated with the first, which, still longer delayed by the additions, was not published until in October of 1883.

The present publication presents mainly studies and observations made in the years 1882 and 1883, although embracing some of those of a later date. During these two years, the agricultural interests of the State did not suffer from insect depredations to an extent equaling that reported for the preceding year. No formidable insect pest presented itself for the first time, as a depredator upon any of the principal field or garden crops, to be added to our already large list of insect enemies. The years, of late, in which such additions have not been made, are unfortunately exceptional ones.

The zebra caterpillar of *Mamestra picta* Harris, was unusually abundant during the autumn of 1883, not only in the State of New York, but also in some of the adjoining States. A severe attack of it on beets (mangolds) was reported by Secretary Harison, of the New York State Agricultural Society, as occurring upon his farm at Morely, St. Lawrence Co. The caterpillars were first observed feeding upon the leaves in September. They continued their destructive work into October, until they had devoured all of the tender leaves and reduced the aggregate of the foliage at least one-half. After this had been done, they attacked the roots, into which they made large surface excavations similar to those eaten by crickets into fallen apples. When the mangolds were taken up on the 15th of October, the caterpillars, as stated by Mr. Harison, were still at work upon them, although some severe

1

frosts (23° and 24° Fahr.) had made them somewhat sluggish in their movements. It was estimated that the yield of the crop had been lessened one-half by the attack of the insect upon it. Quite a number of kinds of garden vegetables were also preyed upon by the Mamestra larvæ, but without committing serious injury. Their favorite garden food-plant seemed to be sweet pea-vines — a long and dense row of which proved very attractive to them. In Massachusetts they have occasionally been extremely destructive to the rutabaga crop. Figures and notice of the insect may be found in Dr. Harris' *Treatise on Insects Injurious to Vegetation*, and in the Report of the Entomologist of the U. S. Department of Agriculture for the year 1883.

The attack upon the pastures and other grass lands in the northern counties of the State by the Vagabond Crambus, *Crambus vulgivagellus* Clem., which was the occasion of so great alarm and of serious loss during the summer of 1881, was not repeated the following year. In accordance with the opinion expressed in my report upon the insect, it proved to be but an exceptional occurrence, resulting from a combination of circumstances which may not again be presented for a long period of time.

The attack of the Corn-worm, *Heliothis armiger* Hübn., upon the corn fields of the southern portion of the State, as a consequence of the unusual heat and drouth of the summer of 1881, was not repeated. Its occurrence within our borders subsequently, has in no instance come to my knowledge.

The injury to clover-seed, from the clover-seed midge, *Cecidomyia leguminicola* Lintn., does not appear to be on the increase in the localities where it had formerly been so abundant, thus affording ground for hope that it may not prove so disastrous to clover culture as it threatened to be, but that the parasites that have attacked it (of which two species are known, *Eurytoma funebris* and *Platygaster error*), are rendering efficient service in its destruction. It seems, however, destined to attain to an extensive distribution, for not only has it reached the extreme western portion of New York, but it has extended into Canada, and excited no little alarm from the amount of clover-seed already destroyed by it. According to the report of the recently appointed Entomologist of the Dominion of Canada, Mr. James Fletcher, the insect is proving even more destructive in its northern extension than in our own State where its operations were first noticed. A few years ago large quantities of Canadian seed were exported to the United States, while at the present not enough is produced for home use. In portions of Ontario, seed clover is reported as an entire failure, as the result of the prevalence of the midge. The first complete destruction of the seed crops in the Dominion, was observed in the year 1883. To the southward of New York the insect is occurring in Pennsylvania and Virginia.

The punctured clover-leaf weevil, *Phytonomus punctatus* (Fabr.), has also steadily and rapidly extended its area of operations from Yates county where it was at first observed in 1881, northwardly to Lake Ontario, and westwardly, to and beyond Niagara river. In August of 1884, large numbers of the beetle carried in their flight by severe easterly winds, were observed on the sidewalks, in the streets, and on house-tops in Buffalo. Thousands were crushed under foot on the pavements by pedestrians. Many were borne westward and dropped into Lake Erie, from the surface of which they were blown upon the shore where they were found in millions, unfortunately not dead, but soon recovering from the effects of their exposure in the water. Many at the same time were observed upon sidewalks, fences and trees at Ridgeway in Canada, ten miles to the west of Buffalo.

The Colorado potato-beetle, *Doryphora decemlineata* (Say), seems to be diminishing. The experience with it in various portions of the State is, that it has been brought under comparatively easy control. Usually a single application of Paris green made early in the season for the destruction of the first brood is all that is required. The diminution in its numbers is believed to be largely owing to the increase of the insects and other animals that feed upon it. In several localities, species not previously known to attack it, are reported from time to time, as making it their prey, as they come to find it convenient and agreeable for food.

The same decrease in its ravages has been noticed in other States. In Pennsylvania and Ohio, according to a statement made by Prof. E. W. Claypole (*American Naturalist*, xvii, 1883, pp. 1174–1175), there was no second brood in 1882, and in the following year, it almost failed to appear at all. Hot and dry summers have been found favorable for its increase, and upon the recurrence of such seasons hereafter, especial care may be required in order that the destructive insect shall not be allowed again to obtain the ascendancy.

Severe injuries have been reported from different localities in the State, to grass, from the destruction of its roots, by the grubs of the common May-beetle, *Lachnosterna fusca* (Frohl.). Hitherto this insect has been one which has almost defied all efforts made for its destruction, but from experiments now being conducted in a quite new direction, results are anticipated which will enable us to control its ravages in a large degree.

The unusual occurrence of immense numbers of grasshoppers in midwinter, while the fields were still largely covered with snow, in some of the southern counties of the State, excited apprehension of serious depredations from their multiplication in the spring and summer. Assurance was given that such fears were groundless, and that the premature appearance of the insect — brought out from hibernation by some warm days in February — would only tend to their destruction, and a consequent diminution of their number and of their injuries in the coming season. An account of the grasshopper, *Chimarocephala viridi-fasciata* will be given in the present report.

A notable event in the autumn of 1883, was the discovery for the first time in injurious numbers in the State of New York, of the chinch-bug, *Blissus leucopterus* (Say) — one of the most dreaded of our insect pests. Its attack was made upon clover, and so severe was it, that many acres of meadow were destroyed in the localities where it made its demonstration. Fortunately the fears that were entertained of its continuance and increase in the following years, were not realized. It has, however, shown its ability at any time when favorable conditions are offered it, of entering upon our grass and grain lands and occasioning severe losses. Several pages of this report have therefore been devoted to the consideration of its natural history, habits, its injuries, etc., with the view of presenting such information as will be found of service in meeting its possible future visitations.

In conclusion, the Entomologist desires to express his regret that the figures illustrating this report are not of the number, and some of them, not of the character, needed for its proper illustration. Under the existing arrangements for the printing of the State reports, the number and the style of execution of the figures that may be introduced, are under the control of the State Printers. To this reason it is owing that the original figures contained in the present report are comparatively few. The following have been engraved for it: Figs. 3, 6a, 11, 12, 13, 14, 16, 21, 22, 23, 25, 26, 34, 36, 40, 41, 42, 43, 50, 52, 53, 55, 56, 57. The remaining ones are from electrotypes obtained from various sources, most of which are indicated in the figures and need not, therefore, be separately mentioned. Many of them will be familiar to entomologists, who do not need their aid; but they cannot but prove new and serviceable to a large portion of our agricultural community, for whom this report is specially prepared. Many years must elapse before good figures of any of our common and more destructive insect pests can be repeated so often that a general familiarity with them and the species that they represent in nature, shall render their further repetition useless.

Respectfully submitted,

J. A. LINTNER.

OFFICE OF THE STATE ENTOMOLOGIST, }
   ALBANY, *March* 10, 1885.   }

NOTES OF VARIOUS INSECT ATTACKS.

A Saw-fly Attack upon Pear-trees.

Mr. S. D. Willard, of the Hammond Nurseries, Geneva, N. Y., sends, under date of May 29th, examples of a saw-fly, the larvæ of which were proving very injurious to the tender leaves of young pear-trees, just as they were commencing their growth. The examples were in very poor condition, having been almost destroyed in their capture. They were sent to Mr. E. Norton, of Farmington, Conn., in the hope that they would be recognized by him, but he was only able to refer them, with some doubt, to *Pristophora*, allied to *grossulariæ*, but probably an undescribed species. Some of the young larvæ were subsequently received, but they came in too poor condition to permit of their being carried to maturity. A later request for additional larvæ (June 12th), could not be complied with, as they had all been killed by an application of hellebore.

Peach-tree Attacks.

A correspondent sends the following communication containing inquiries of insects attacking his peach-trees:

Is there no remedy against the peach-tree borer except getting it out of the roots with a wire? I have been caused much trouble by this pest, the trunks of my peach-trees being terribly cut and scarified from just under the earth down into the roots. It is laborious in the extreme to cut them out two or three times a year; for it is next to impossible to get them all at the first trial. How would an application of coal-oil do? Would it kill the tree? Will ashes spread around keep the insect from laying the eggs?
In addition to the grub or borer, which is a fat white worm, half an inch in length, with a brown head, and which is found snugly ensconced under the bark after being followed through his various meanderings, I find numbers of fine, white, thread-like worms, of all lengths, from a quarter of an inch to an inch. They also seem to be doing injury, but are found behind the grub in the soft sticky wood. I also find the common earth-worm imbedded in this part of the tree, but think that the latter may be attracted by the moisture caused by the sap exuding from the wounds, and intends no harm. Am I right? Is not the worm found in peaches and caused by curculio, almost exactly like this borer of the bark? Please be kind enough to tell me something of the habits of the two beetles, when the eggs are laid, etc.

Can you tell me also, what causes the fungoid growth upon the pieces of the peach-tree sent ?

In addition to the most effectual means of destroying the peach-tree borer, *Ægeria exitiosa* Say, viz., that of digging them out of their burrows which is regarded by many peach growers as not a laborious or difficult operation — the application of hot water to the base of the tree has been recommended, and in many cases has proved quite successful.

Mounding to prevent the peach-tree borer. — The trees of an orchard having been once thoroughly "wormed," preventive measures should be resorted to, that they may not be restocked with the borers. There is good evidence that this is practicable in all but quite young trees by the method known as *mounding*. It is simply the throwing up around the base of the tree, at any time before the parent moths are abroad for the deposit of their eggs (usually in the month of July,) a mound of earth of about a foot in height, pressed by the foot closely to the tree. In the following years a few inches of earth may be added annually. By this means, the roots of the tree where they are given off from the trunk, are placed out of reach of the insect. The mounding is believed also to have a beneficial influence on the health of the trees in prolonging their period of bearing and exempting them from disease. Several extensive peach growers claim for this method that it has given them entire exemption from the ravages of the borer, at the cost of a very little labor — one man being able to mound fifty trees in a day. A mound of ashes might prove quite as effectual as one of earth.

Coal-oil application not safe. — The application of coal-oil to the trunk of the tree, which is suggested, would be a hazardous experiment, unless first tried upon a tree or two, for it has been known not unfrequently to kill the trees that have been treated with it. At certain seasons of the year, coal-oil has been applied to the bark of trees for killing scale insects, with scarcely any harm resulting from its use.

Dipterous larvæ. — The " fine, white, thread-like worms," associated with the borers in their burrows, are the larvæ of some species of fly, which have not, so far as we know, been carried to their final stage. They are harmless scavengers, as they feed upon the exuding sap and gummy matter. Some of the smaller myriapods, or hundred-legged worms also frequent the decayed bark and wood of the injured peach-trees. The earth-worms observed, as supposed, are not injurious.

The curculio in the peach. — " The worm found in the peach and caused by the curculio," is very unlike the peach-tree borer, it being the larva of a beetle, *Conotrachelus nenuphar* (Herbst), appearing as a yellowish-white footless grub, with a yellow head and two pale lateral lines on its body; its length at maturity is about one-third of an inch. The peach-tree borer is the larva of a moth, and is a caterpillar (not a grub) with sixteen legs. When full-grown it exceeds a half-inch in length. The moth deposits its eggs in the crevices of the bark just above the surface of the ground. The curculio beetle cuts with its beak a crescent-shaped incision in the plum into which it inserts its egg. It will thus be seen that the appearance and habits of the two insects are widely different.

The fungus on the peach twigs. — The fungoid growth upon the peach twigs is the common fungus, *Stereum complicatum*, as determined by the State Botanist, Prof. C. H. Peck. This fungus is by no means peculiar to the peach, but is common everywhere on dead and decaying wood, and does no injury to living plants.

APPLE-TREE ATTACK BY THE CANKER-WORM.

Inquiries similar to the following are frequently received from various parts of the United States, particularly from the New England States, and with increasing frequency from localities in the State of New York:

I send you by to-day's mail a sample of worms I find in my apple orchard, suspended by a single thread from the branches. Are they army-worms, and what is the best way to destroy them? There are more this year than last, and the orchard of 2,000 trees presents a brown appearance after they have fed on the leaves for a while. (W. E., Carrolton, Ky., April 28.)

The caterpillar sent is the spring canker-worm, *Anisopteryx vernata*, of which, notwithstanding so much has been written, many persons are yet entirely ignorant, from the fact that although the species has a very extensive distribution, yet it is quite local in its appearance, prevailing, perhaps, in one or two counties of a State, or it may be in one or two orchards in a locality. This is owing to the fact that the female is wingless, and, therefore, its distribution must usually depend upon assistance afforded it in passing from one orchard, or an isolated tree, to another.

Extensive range of the canker-worm. — It is capable of becoming a great pest — one of the greatest with which the orchardist has to contend. Its distribution is from Maine to Texas, and it has been very destructive in Massachusetts, Illinois, Missouri, Wisconsin, Ohio, and in some other States.

Birds that prey upon it. — Where birds have been protected, they are often the means of preventing its great increase, for several species prey eagerly upon it, and during the time of its prevalence find in it by far the larger portion of their food. The cedar bird (*Ampelis cedrorum*), devours large numbers, and a flock of them in an orchard can render most excellent service in their destruction. As many as one hundred of the caterpillars have been found in the stomach of one of these birds. Next in usefulness come the indigo bird (*Passerina cyanea*), the chickadee (*Parus atricapillus*), the black-billed cuckoo (*Coccygus erythropthalmus*), the summer warbler (*Dendraeca æstiva*), the rose-breasted grosbeak (*Goniaphea ludoviciana*),the blue bird (*Sialia sialis*), the king bird (*Tyrannus carolinensis*) and the robin (*Merula migratoria*), in the order mentioned, as shown by examination of the contents of their stomachs by Prof. Forbes.

Climbing habit of the female moth. — It is not safe, however, in the presence of so dangerous an enemy, to commit our apple-trees to the care of the birds; but earnest efforts should be made to confine the attack to the narrowest possible limits, and to arrest it as soon as possible. By far the best method to employ with this insect pest is prevention. The wingless moths emerge from their pupæ in early spring, buried a few inches in the ground about the trunk of the tree in quite a limited space — the larger portion in a circle, the radius of which would not exceed six feet. As soon as extricated from the pupæ, the moths direct their course toward the tree, to ascend the trunk, and deposit their burden of eggs in the crevices of the rough bark of the upper portion of the trunk.

Tar bands for arresting the moth. — *The ascent of the trunk should not be permitted*, for it can be prevented with comparatively little labor. The method adopted with excellent success for many years was that of encircling the trunks a short distance above the ground, with a band of heavy cloth, to be covered with tar, or better, tar and molasses; to be renewed every two or three days, or whenever it becomes so hardened by exposure as no longer to serve the purpose of fastening the moths which attempt to cross it in their ascent.

Tin-band protector. — A still better method than this has since been devised, and is thus described by Dr. Le Baron in his second report as State Entomologist of Illinois: "Take a piece of inch rope — old, worn-out rope is just as good as new; tack one end to the trunk, two feet or less from the ground, with a shingle-nail driven in so that the head shall not project beyond the level of the rope; bring the rope around the tree, and let it lap by the beginning an inch or two; cut it off and fasten it in the same manner. Get the tinman to cut up some sheets of tin into strips four inches wide, and fasten them together endwise, so that they shall be long enough to go round the trees over the rope band, having the rope at the middle. Let the ends of the tin lap a little; punch a hole through them and fasten them with a nail driven through the tin and rope into the tree."

The operation of this is simple. The moths ascend the tree to the rope and here congregate, for it appears to be entirely contrary to their nature to travel downward. When the space between the tree and trunk has become filled with them, then the new-comers may pass over their compacted bodies, and by dint of effort ascend the outside of the tin; but reaching its upper edge, observation shows that they will not descend upon the inside, but travel round and round the upper edge until they abandon the attempt to reach the tree, and fall to the ground.

Other methods have been used for preventing the ascent of the moths but none seem to be so simple and effectual as the above. A somewhat similar one, upon the same general principle, is the following:

An improved tin protector. — Take a strip of tin four inches wide of

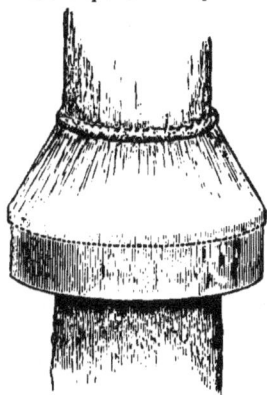

sufficient length when encircling the tree to leave a space of about six inches. The upper edge of the tin is bent over so as to receive beneath it a piece of muslin as long as the tin and eight inches wide, to be held in place by pounding down the tin. The ends of the tin are bent in opposite directions so that they can be hooked together. Placing this around the tree with the cloth upward, the cloth is to be firmly bound to the tree by a strong cord. This method is shown in the accompanying illustration. In either of the above methods, the eggs which will be deposited in large numbers below the obstruction, may be easily killed by brushing them with

FIG. 1. The suspended tin-band tree protector for preventing the ascent of the Canker-worm moth.

kerosene oil, without injury to the tree, unless an excessive quantity (a very little is needed) should be used.

Killing the larva. — When, as in the inquiry above made, the moths have not been prevented from ascending the tree and depositing their eggs, and the caterpillars have hatched out and are rapidly destroying the foliage, then there are two good methods for their destruction which may be resorted to.

Spread a thin covering of straw under the tree as far as the branches extend, and set fire to it, at the same time jarring the tree, and as the caterpillars drop by their threads, sweep them down with a pole into the fire.

Apply Paris green in water to the trees with a force-pump, in the proportions and manner probably known to every orchardist. It may be used with perfect safety upon apple-trees so early in the season as the time during which the canker-worm prevails.

2

A GOOSEBERRY FRUIT-WORM.

A gentleman from Delhi, N. Y., writes:

I find my gooseberries dropping badly, and have experienced the same trouble for the two years past. Upon examination for the cause, I find that nearly every berry contains a small white worm between the skin and the pulp, which sometimes has worked its way into a seed. I send some of the fruit picked up from under the bushes, and think that you will find the worm in each. The worms are not more than one-eighth of an inch in length. Can you suggest a remedy for next year, as I presume the work is done for this season?

A portion only of the gooseberries sent contained the "worm"—those which were discolored and shriveled. In the others, although each bore a scar as if an egg had been deposited, no living thing could be found.

The depredator is the larva of a lepidopterous insect — a small moth, and in all probability that described by Dr. Packard as *Pempelia grossulariæ*, and more lately known as *Dakruma convolutella* (Hübn.). At this stage of growth, it is impossible to identify positively the larva from the description published of the adult form. Later, if it be found that when the interior of the berry has been consumed, the larva passes into an adjoining one through a silken tube with which it connects the two, then scarcely a doubt will remain of its being the above-named species which was first detected by Dr. Fitch, in the State of New York, about the year 1855. Its natural history was worked out by Mr. Saunders, of London, Ontario, ten years later, and from material furnished by him, it was described and named by Dr. Packard, in his *Guide to the Study of Insects*.

Mr. Saunders states that the moth appears abroad in the month of April, and is ready to deposit its eggs as soon as the fruit is well formed. The larva hatches and usually completes its growth about the middle of June, when it lowers itself to the ground, and constructs its cocoon among the leaves or in the superficial soil. Its pupation continues through the winter, the moth making its appearance in the early spring, as above stated.

If care be taken to pick from the bushes all the prematurely ripening and shriveled fruit, and all that drops to the ground, this insect need not prove a serious pest. One caterpillar allowed to mature, may the following season be the progenitor of fifty or more depredators upon the same bushes.

Dusting the bushes with fresh air-slacked lime about the time when the eggs are deposited — the last of April or first of May — is recommended, and has been thought to be attended with beneficial results in keeping away the moth.

INSECTS AND FUNGUS ON QUINCES.

A package of nearly half-grown quinces was received from Union Springs, N. Y., under date of July 27, with the following inquiry of the cause of their appearance and condition :

I have just received from C. E. Cook, of South Byron, N. Y.—a large orchardist, specimens of quinces containing an insect, and singularly infested with a rusty fungus, which I forward to you for examination. Mr. C. thinks it is all caused by the insect inside the fruit. I am informed that much of the fruit is thus attacked, and it is feared that the difficulty may become a formidable one. What is the insect, and is there any remedy ?

The Quince curculio and the Apple-worm.— The quinces contained two insects burrowing within them, viz., the larva of the quince curculio, *Conotrachelus crataegi* Walsh, and the larva of the codling-moth, *Carpocapsa pomonella* (Linn.). The former was the most numerous, as in some of the quinces examined, four examples of it were found. They are easily to be distinguished from the apple-worm, as they are without feet. Walsh has described them as of " an average length when full-grown, of o.32 of an inch, four and one-half times as long as wide, straight, opaque-whitish with a narrow, dusky dorsal line generally obsolete on the thorax, and a few very short hairs; distinct lateral tubercles on all the joints; head rufous, mandibles black, except at base, and distinctly two-teethed at tip."

The beetle is figured in the Third Missouri Report, as bearing only a family resemblance to the plum-weevil, *Conotrachelus nenuphar* (Herbst), being larger, with a longer beak or snout, and a body which is broadest across the base of the wing-covers.

Food-habits of the Quince curculio. — It was first found feeding upon the fruit of the black thorn, *Crataegus tomentosa*, and from this food-plant its specific name was taken. In the Western States it occurs more frequently in this fruit, but in the east it more commonly infests the quince. Instead of the crescent cut characteristic of the plum-weevil, it simply makes a puncture for the reception of its egg. Professor Riley states that the larvae work, for the most part, near the surface, and do not enter the heart of the quince, but in those examined by me, they had penetrated quite to the interior and were feeding upon the seeds.

Its transformations. — After about a month's feeding, the larvae desert the fruit and burrow into the ground, where they remain unchanged from the larval state throughout the winter and until early in May, when they transform to pupae and shortly after that to the perfect insects.

Its injuries. — This insect has been known to prove very destructive to quinces. Dr. Trimble records an instance in which, in the latter part of October, five or six hundred of the larvæ were taken from the bottoms of two barrels of quinces gathered but four days before. In an orchard, in New Jersey, in 1870, of nearly three hundred trees, scarcely a single quince could be found free from the attack of the insect. The injury was estimated at about $800.

Remedies. — The remedies against this insect are, first, the jarring method, which has proved so successful with the plum weevil, for the quince weevil falls quite as readily at a sudden jar. The proper time for employing this remedy may be found by experiment, during the months of June, July and August. Second, picking the infested fruit and destroying it. The presence of the larvæ within the fruit may usually be discovered by the black grains of its excrements attached to its surface. These rejecta are easily discernible when held in place by a fibrous fungus growth which is often associated with the insect attack. Mr. Cook informs me that the infested fruit, unlike that attacked by the apple-worm, remains firmly attached to the tree. Its collection, therefore, involves more labor than the simple gathering of the fallen fruit.

The apple-worms in the quinces. — The apple-worms were not as numerous as the curculio larvæ in the fruit sent for examination, and in no case did I discover more than one in a quince. They displayed a greater readiness, even when apparently not mature, to leave the fruit in the box in which it had been sent to me, than the other larvæ. Possibly even the smaller ones may have been mature, and all may have been seeking retreats for their pupal change.

A fungus attack. — All of the fruit received was also infested with a fungus. Some of the quinces had almost their entire surface covered with it, and its yellow growth extending quite a little distance into the fruit. The fungus has been described and named by Prof. C. H. Peck, N. Y. State Botanist, as *Rœstelia aurantiaca* — the specific name having reference to the rich golden color of its spores. Upon inquiring of Prof. Peck if the fungus was peculiar to the quince, and to what extent it had been observed to be associated with insect attack, he replied: " It occurs also on unripe fruit of *Amelanchier Canadensis* and of species of *Cratægus.* I do not think that it is always or necessarily accompanied by an insect attack, although of two specimens of quinces recently received from Rochester, and bearing the fungus, both had larvæ in them. In the Amelanchier fruit, I know that the fungus sometimes, at least, works alone." [See also *the Country Gentleman,* for Dec. 17, 1885, p. 1016.]

In answer to inquiries made of Mr. Cook of this fungus attack, the following reply was returned, under date of August 3d:

I send by mail, another package of the infested fruit, as desired.

In reply to your questions: 1. The fungus was first noticed this year.
2. I think that the fungus is found without the insect burrows.   3. The
fruit does not fall when infested, but adheres firmly to the tree.   4. I
find them in great numbers in all stages of infection.   I have several
hundred trees bearing, from which I could pick several bushels of diseased fruit.   5. I send inclosed several affected twigs, independent of
fruit.   You may observe that it does not affect the foliage, there being
even perfect leaves in the fungus growth.   It is quite serious, as it frequently continues until all the foliage is almost entirely ruined."

The quince twigs sent by Mr. Cook were affected by the same fungus
that had attacked the fruit, in some examples showing as a thick irregular swelling upon one side of a twig, and in others completely surrounding it for an extent usually of about an inch.   The larvæ had
also attacked these fungus swellings, and had burrowed them with their
channels, disclosing their presence and their operations by their excrementa adhering to the surface.

A fungus-feeding fly. — A large number of minute larvæ of some
species of fly were associated with the fungus, feeding upon the orange-
colored spores.   Their maximum length was about one-tenth of an inch.
They were pointed at both ends, but a little more acutely at the head.
Their color was an orange nearly as bright as their food.   Some had
already transformed to the pupal stage, and a few flies which escaped
from the box when it was first opened, were in all probability, the newly
emerged insects.   None of them were secured for examination.   The
species has probably not been named, for of the number of Diptera
which are known as spore-feeders, but few have received study.

Remedy for the fungus.— The only means, so far as we know, for preventing the spread of this fungus, is to remove and destroy by burning or
otherwise, every twig or quince giving indication of the attack.   It is
possible that dusting the tree with flowers of sulphur might be of service in arresting the disease.

THE PLUM WEEVIL ATTACKING APPLES.

Apples bearing the distinctive crescent mark of *Conotrachelus nenuphar*
and quite badly scarred with them, were received June 26, 1882, from
Mr. H. J. Foster, of East Palmyra, N. Y.   The fruit was shriveled and
discolored on one side, apparently from several contiguous punctures of
the weevil made for the purpose of feeding.   Upon cutting into the
apples, the young larvæ of the weevil were found.

On making inquiry of Mr. Foster if the attack of the plum curculio upon the apples in his vicinity might not be the result of the scarcity
of their natural food, he informed me that there were very few plum

trees in the neighborhood, and that most of the farmers were compelled to purchase plums for their family use, from a locality where they were abundant, on "the ridge" or near Lake Ontario ten miles to the northward. The curculio attack had been increasing for several years, and was quite serious the preceding year. Almost every farm of one hundred acres had from five to twenty acres of apple orchard, kept in permanent pasture, yet but little fruit could survive the combined attacks of the codling-moth, the canker-worm, the tent-caterpillar, the curculio, et cet., unless stock was kept in the orchards to feed the grass closely and pick up the fallen fruit.

The earliest attacked apples, fell to the ground with the contained curculio larvæ. The later ones, it was thought, remained upon the trees.

Upon the 8th of July, Prof. Charles H. Peck, sent to me from Petersburg, Rensselaer county, N. Y., examples of the plum weevil, *Conotrachelus nenuphar*, with specimens of the apples in which he found it ovipositing. In some cases, it had oviposited in nearly all the apples upon a tree, causing many of them to fall to the ground.

PHYTONOMUS PUNCTATUS FEEDING ON BEANS.

Examples of this beetle were brought to me on July 2d, by Mr. J. F. Rose, of South Byron, Genesee Co., N. Y., which had been given to him by a farmer in that town with the statement that they were feeding in large numbers, upon the leaves of some field beans, and were rapidly destroying the crop. Mr. Rose knew the beetles to be abundant in localities in the town, and that they had been quite destructive to clover, for during the month of June of last year (1883), he had seen in a clover field that had been turned under for fertilizing, the larvæ so numerous on the surface of the ground the morning after the plowing that he was able to count fifty within the area of a square foot.

As the beetle had not been recorded as feeding upon beans, it was desirable that this first report of a new food-plant should be verified, before accepting it; yet as the clover and the bean both belong to the *Leguminosæ*, it did not seem improbable that with clover not convenient of access, the insect might transfer itself to some other genus of the order.

In compliance with the request made, Mr. Rose, upon his return home visited the field in which the beetles were reported as eating the bean leaves, and found them actively engaged in the work. Under date of July 14th, he wrote as follows :

I send you by this mail specimens of the beans eaten by the clover-leaf beetle together with some of the beetles taken from the beans. Mr.

White, from whose farm they were taken, as were also those handed to you in Albany, had been over the field and Paris-greened it as he would have done for potatoes. He had found as many as twenty-five beetles upon a single hill, and he thinks had he not used the Paris green, not a leaf would have been left. The poison did its work effectually and at this time but few of the beetles are to be found. The bean field was adjoining a field of clover which was plowed under, and the beetles did not show themselves on the beans until after the plowing. Many other fields of beans are reported as infested, although the one of Mr. White is the only one that I have visited.

The plants sent to me gave evidence of the possible injuries of the weevil when its attack is made in force, for almost every leaf had been entirely consumed. Much of the feeding had in all probability been done by the larvæ.

Wishing to see the method of eating in the beetles received, some fresh leaves of bean and some pods procured in the market were given them. They declined to eat the leaves, even when confined with them alone for several days, but they fed greedily upon the pods, excavating holes in different portions, of perhaps a tenth of an inch in diameter.

THE "FRENCHING" OF CORN.

A correspondent from Rock Hall, Md., sends a specimen of what, in his vicinity, is known as "frenchy corn," and asks if it is the result of an insect attack, or of conditions existing in the soil. It sometimes prevails to such an extent as to destroy an entire crop.

What is understood by "frenchy."— The term "frenchy" is one used in many localities in the Southern States in quite a general way, and is indifferently applied to diseases (aggravated form of measles, small-pox and other contagions), to various insect injuries to vegetation, and to diseased conditions of animal or vegetable life not understood. It is, however, more frequently used in connection with corn than any other plant. According to newspaper reports, thousands of acres of corn are annually destroyed in the Southern States through "frenching."

"Frenching" resulting from insect attack.— The young corn-plants sent to me, having a growth of two feet in height, had lost their tap-root, and showed clearly in their slenderness and other conditions, an impaired growth. Throughout their entire length were numerous punctures, many of which were of the size and shape of pin-holes, while others were larger and of an oval form, and had evidently increased in size and changed in form through the process of growth of the leaves. A careful examination left no room for doubt that these punctures had been made in the unfolded leaves of the plant by the beak of one of the snout beetles — *Curculionidæ*.

The insect probably Sphenophorus sculptilis. — The locality from which the material was received rendered it highly probable that the species was the sculptured corn-curculio, *Sphenophorus sculptilis* Uhler.[*]

This pernicious beetle extends over a large portion of the United States, occurring in Georgia, Kansas, Missouri, Illinois, and some of the Eastern States. It is particularly injurious, at times, in New York. It is of a small size, black, about three-tenths of an inch long, subcylindrical or a long-oval, its wing-covers marked with rows of punctures, its strongly curved beak of the thickness of a stout horse-hair and about one-third the length of the body. In the month of June the beetles may be found just beneath the surface of the ground, with their beak thrust into the young stalks from which they are drawing the sap. Through this method of attack the little holes above mentioned are made, and the growth of the plant arrested.

The attack and how to prevent it. — It is probable that the eggs of the beetle are deposited in the holes made in the stalk by its beak. The larva upon hatching from the egg, burrows downward and destroys the tap-root. When full-grown, it changes to the pupa state within the lower portion of the stalk. It is thought that it transforms to the perfect state — the beetle, during the autumn, hibernating in this stage, and coming abroad the following May or June to deposit its eggs. When the attack of this insect is first noticed, if sand is well moistened with kerosene oil, and a small handful distributed among the young blades of corn in a hill, as the oil is carried into the soil by the rain, it should kill the beetles engaged in their depredations. The same dressing applied earlier may be more effectual by preventing, through its odor, the deposit of the eggs. If only a few hills show the attack, the ground might be drawn away by hand from the stalks, and the beetles taken from them and destroyed.

THE BED-BUG INFESTING A LIBRARY.

It is not very often that this insect occurs under the conditions mentioned in the following communication — among books and papers — where it would be so difficult to destroy it:

Will you tell us something about the bed-bug, what its habits are, when it "spawns," what it eats, how long it lives, and if it ever dies? I ask because I have moved into a house that I find was already occupied by several colonies of the pest. The room in which I have my library has the most. They are in my files of papers and periodicals. They seem to grow fatter every day, but for the life of me, I cannot tell what they live on. The beds and furniture are poisoned with cor-

[*] See First Report on the Injurious and other Insects of the State of New York, 1882, pp. 253-263.

rosive sublimate, and are free from them. Can it be that they live on
the paste on the wall paper ? As for remedies, or rather exterminators,
I have used two — the sublimate and red pepper. The latter, I have
sifted through my papers and books, and wherever I could get it; but
instead of driving them off, they seem to fatten on it; it is a sort of
condiment that helps them to relish their dinner the better. If you can
tell me what to do, and answer the first question also, you will do me a
great favor. I would move to another house, but there is not a vacant
one in the place.

The bed-bug is generally believed to be a native of America, and its
prevalence in Europe to have resulted from its introduction into that
country in the wood imported from America for the rebuilding of
London after the great fire of 1666. This, however, must be an error,
for although it was not common in England, " it was well known in
some parts of Europe before that time, and is mentioned by Disscori-
des." An old writer records it in Europe as early as the year 1503, and
there is reason to believe that it was known to Pliny and Aristotle. It
was first described and named by Linnæus as *Cimex lectularius* — its
specific name denoting its pertaining to the bed. At present, it is
known as *Acanthia lectularia.*

I am unable to find a full statement of its natural history. It has
been treated of at considerable length by some of the old European
writers to whose papers I have not access. Unfortunately our modern
authors, in their references to the unsavory subject, dismiss it with a
few cursory remarks and the statement "that its habits are too well
known to call for further notice." There are, therefore, several points
in its domestic economy of which we are still in ignorance.

The eggs are deposited, it is said, in March and April. They are of
an oval form, white, and open by a little lid to give forth the occupant.
A period of eleven weeks, under ordinary circumstances, carries them
to their maturity. Where they can obtain the amount of food needed
for their development, in an unoccupied apartment, is a question that
cannot be satisfactorily answered. It is possible that in rooms where
the walls are damp, the moistened paste of the wall-paper might serve
them for nutriment, for it must be remembered that their food is taken
by suction through a tube, and not by means of biting jaws. As quite
a number of the Hemiptera are known, when pressed by hunger, to
prey upon one another, it is not improbable that the young of this species may find it convenient to expedite the death of their aged and in-
firm parents. And possibly, they may feed upon the juices of flies and
other insects, for they are known not to confine themselves entirely to
the human species, but to infest, occasionally, chicken-coops, and also,
it is stated, dove-cotes. When they have reached maturity, they are

3

capable of living a long time without food. De Geer, a distinguished
Swedish naturalist of the last century, kept some specimens alive in a
sealed bottle for more than a year without food.

Probably the best method that could be adopted by the inquirer to free
his library and other apartments from so serious an invasion of this pest,
would be to fumigate with brimstone. Houses which after standing
long unoccupied have been found swarming with the bugs, have been
effectually freed from them by this means. Place in the center of the
room a dish containing about four ounces of brimstone, within a larger
vessel, so that the possible overflowing of the burning mass may not injure the carpet or set fire to the floor. After removing from the room
all such metallic surfaces as might be affected by the fumes, close every
aperture, even the key-holes, and set fire to the brimstone. When four
or five hours have elapsed, the room may be entered and the windows
opened for a thorough airing.

PSYLLA BUXI UPON BOX, AT WEST FARMS, N. Y.

Mr. James Angus, of West Farms, N. Y., sent to me, under date of
May 23d, a small insect upon box, *Buxus sempervirens*, which according
to his statement, entirely covers the plants at this season of the year.
They secrete a white floculent matter, and cover the leaves with a thick
honey-dew. Later in the year their gummy exuviæ greatly mar the
beauty of the plant. The only remedy thus far found against them
has been to beat the plants violently with a broom and then to rub the
dislodged insects into the ground. This was done several times during
the season.

The insects were found to be *Psyllidæ* — probably of the genus *Psylla*,
but as they were at this time in their pupal stage, no determination could
be made. They averaged one-tenth of an inch long, of an apple-green
color, head broad, nearly as broad as the thorax; thorax and abdomen
about equal in length and breadth; antennæ apparently four-jointed,
the last joint black and the others black at the tip.

They were confined in a small jar with the leaves of the box, and on
the 29th of May the first imago was disclosed. Upon submitting specimens to Prof. Uhler for determination, he kindly returned the following
reply :

I have carefully compared your specimens of *Psylla* with *Ps. buxi*,
of Europe, in my collection, and I fail to discover any differences. As
the wings of one specimen are clouded, I take it for granted that this
one is immature; the others have transparent wings. *P. buxi* Linn.,
is described in *Syst. Nat.*, ii, p. 738; *Fabr. Spec. Ins.*, ii, p. 391; *Reaum.
Ins.* iii, pl. 19, figs. 1–14 ; *Foerster, Verhandl. Nat. Verein. Preuss.
Rheinl.*, 1848, iii, p. 71, No. 3. It is common in England and Germany.

This occurrence of the species at West Farms is of no little interest, as it had never before been reported in the United States.

In the report of the Commissioner of Agriculture for the year 1884, at page 410, Prof. Riley has published a brief notice of this discovery.

An Aphis Attack on Roots of Peach-trees.

Notice of an aphis attack upon the roots of seedling peaches has been communicated to me by Mr. Lorin Blodget, of Philadelphia, Pa. The injury to the trees was first noticed in the year 1881, but its cause remained unknown until the early part of July, 1884, when upon pulling up a seedling peach-tree just beginning to wilt, its stem for an inch below the surface was found to be crowded with dark colored aphids: numbers of ants were associated with them. In following up this discovery — of a hundred trees examined, one-half at least were found so seriously injured that they were past recovery and were accordingly destroyed. It was doubtful if any of the remainder could survive the attack. In one instance, some aphids were discovered above ground, upon the succulent shoots about a foot long, of a three-year-old tree, which were densely crowded with them, presenting "a singular sight, with their black, shining backs, covered with ants and with large flies often upon them." During forty years' growth of seedling peaches, no injury of this character had been observed before this attack.

Some examples of the insect sent to me, were dead when received and in otherwise poor condition. They were wingless, entirely black, although said when living to be brownish-black. The species evidently belonged to the genus *Myzus*, and probably (the material being too poor for positive determination) to *persicæ* of Sulzer. A letter to Mr. Blodget expressing this belief, was subsequently handed by him for publication to the *Gardener's Monthly and Horticulturist*, for September, 1884 (xxvi, p. 271-2).

Mr. Blodget claims for this species a remarkable power of endurance of cold,— that it lives and thrives all winter. So late as the 25th of October, after a severe frost, a shoot of a peach-tree cut off close to the ground, was entirely black for six inches of its length with the insect in lively condition. This was sent to the Department of Agriculture, at Washington, for examination. Answer was returned that it had been referred to Professor Riley, and the aphis pronounced by him to be *Myzus cerasi* of Fabricus, which he had long known as injurious to peach-trees, especially to young ones in the nursery, by working on the roots.

An earlier notice of the operations of what was probably the same insect upon the roots of peach-trees, was given by Mr. Glover on page 37 in his Report on Homoptera contained in the Report of the Commissioner of Agriculture for the year 1876, pp. 24–46.   It is this:

Colonel Wilkins, of Riverside, near Chestertown, Md., a very exten-sive peach grower, last spring wrote to the Department of Agriculture that an aphis or plant-louse similar to those infesting his peach-tree leaves was at work on the roots also, and was killing them by hundreds. Professor P. R. Uhler, of the Peabody library in Baltimore, to whom Colonel Wilkins applied, visited the infested peach orchards, and found the statement to be perfectly correct, and that an underground aphis or plant-louse, not differing from those on the leaves, was doing immense injury to the young trees by sucking out the sap.   Professor Uhler also stated that both insects are different from the *Aphis persicæ* above men-tioned, and probably is a new species, closely allied to, if not identical with, the *Aphis chrysanthemi* of Europe.   The insects on both roots and leaves were about 0.08 of an inch in length, with the contour of a broad Florence flask, of a blackish-brown color, and the two varieties could not be distinguished from each other when placed side by side.

Dr. J. C. Neal, in his Report to the Entomologist of the Depart-ment of Agriculture at Washington, of insect injuries observed by him in Florida, in 1882, mentions the occurrence of "lice on roots of the peach," which may also have been identical with the above (*Bulletin* No. I, *U. S. Dept. Agricul.—Divis. of Entomol.*, 1883, p. 36).

In some correspondence with Prof. Riley in relation to this species, he pronounces it identical with forms that he had received in past years from others, and especially in April, 1875, from E. Wilkins, of Chester-town, Md.   It had been very destructive for years on the Atlantic sea-board, and in notices of it published and in replies to correspondents, he had generally referred it, with doubt, to *A.* (*Myzus*) *persicæ* Sulzer; but in consideration of the differences in the descriptions of this species by Boyer, Sulzer and Kaltenbach, and the general characters being so near to *Myzus cerasi* (Fabr.), he had recently referred it rather to this latter species.

Only the wingless form was sent to me by Mr. Blodget.   I was in-formed that the winged form was observed by him to be quite abund-ant during some warm weather in the latter part of November, but he was unable to comply with my request for some examples, as a severe gale, followed by extreme cold weather, had driven them all away or destroyed them.

Whether this should be regarded as a distinct species, or whether it shall prove to be only the root-inhabiting form of *Myzus persicæ*, can only be determined with ample material at hand for study, and careful comparison with the various descriptions of the allied species.   The lat-ter view is apparently held by Professor Uhler, who in his late very valuable contribution to the *Standard Natural History*, published by Cassino & Co., has written:

In many species of the true plant-lice, it is now well established that there are two types of the same insect, the one inhabiting the roots, and living there during the colder part of the year unwinged — the other inhabiting the leaves and twigs throughout the spring and summer. This is notably the case with a small black aphis which injures, and even destroys, the peach-trees of eastern Maryland, Delaware and New Jersey.

Remedial measures. — The following measures, substantially, were recommended by me, to Mr. Blodget, as promising success in arresting the attack of this root-aphis:

Remove a portion of the surface soil from above the roots of the young trees, and apply hot water of as high a temperature as experiment would show could be borne by the trees without injuring them. This should prove fatal to the aphides.

A kerosene and soap emulsion, similarly applied, could hardly fail of proving effectual, as also the Soluble Phenyle noticed in the First Report on the Insects of New York, pp. 48–50. Simply pulling up the infested seedlings and burning them, cannot be relied upon to arrest such an attack, if others are to be grown in the same ground. Numbers of the insects would be left in the soil surrounding the roots, where, after the removal of the peach roots, other food might be found affording nourishment and providing for the continuance of the species. It was formerly believed that the species of Aphides were confined to a single food-plant. Dr. Fitch was of the opinion that *A. cerasi* pertained only to the garden cherry, *Cerasus vulgaris*, and that each " of our native or wild cherry trees [five species] have plant-lice peculiar to them which seldom if ever fix themselves upon the foliage of the other kinds ; " yet the *A. cerasi* (*Myzus cerasi*) has since been found upon the plum and the peach, and perhaps upon so different a food-plant as the *Chrysanthemum*, Prof. Uhler having expressed the opinion that it may be identical with *Aphis chrysanthemi.* If then, the attack upon the seedlings be not entirely arrested by the measures above recommended, a new locality should be selected for growing them, and the ground that they had occupied treated with a liberal application of fresh gas-lime, which, when washed in by rains, could be relied upon for killing the Aphides at the moderate depth at which they would occur.

When the roots of larger trees, which are too valuable to sacrifice, are attacked by this insect, the method might be used which was found highly successful in France for the destruction of the *Phylloxera* upon the roots of grapevines, viz., the introduction of bisulphide of carbon into the soil by pouring a small quantity of it into a hole two or three feet in depth. made in the ground by a pointed bar, confining it therein by packing the ground over it so that in its gradual decomposition it

may permeate the soil, and with its poisonous vapor kill the Aphis (see
1st *Report Insects of New York*, page 47). The "nether-inserter" of
Dr. Barnard, noticed in a following page, could advantageously be used
for the conveyance of this powerful insecticide into the ground.

As superior to the above, recommendation has been made by M. Du-
mas, Permanent Secretary of the Academy of Science of France, for the
arrest of the grape Phylloxera, of the sulpho-carbonates of potassium
and sodium, and of barium. They would be equally serviceable against
all root-inhabiting aphides.

The sulpho-carbonate of barium decomposes under the influence of
carbonic acid, and evolves sulphuretted hydrogen and bisulphide of carbon. Placed in the ground, by its slow decomposition, it should prove
a powerful insecticide.

Sulpho-carbonate of potassium, in addition to its toxic effect, has also
a direct invigorating influence upon the plant.

The use of these sulpho-carbonates was suggested by the need of
some substance that would evaporate less quickly than the bisulphide of
carbon, and thereby infect with its vapors all the surrounding soil.
They should be reduced to fine powders and spread over the surface of
the ground before the heavy autumnal rains.

PULVINARIA INNUMERABILIS UPON GRAPEVINES.

N. C. Scudder, M. D., of Rome, N. Y., in sending examples of *Pul-
vinaria innumerabilis* Rathvon, under date of July 20, 1884, writes: " I
send specimens that I took from a grapevine in my garden. The vine
was very thrifty and had an abundance of small clusters of grapes. In
the latter part of June I noticed that the grapes began to wither and
dry up. On searching for the cause, I found these specimens fastened
to the vine, from the surface of the soil upward for about three feet, the
greatest number being gathered about the enlargements, or where the
vine divides into smaller branches. I searched diligently among upper
branches, but found none. In all cases I found them attached *under* the
loose bark. What I took to be the eggs, I found to be all alive,
and when I emptied the contents of the shell into my hand they began to
crawl. * * * * Since I began destroying these insects, what few
grapes were left'have grown and shown no signs of destruction as yet."

In acknowledging the above, in addition to the removal of the scales,
recommendation was made of application at the time of the hatching
and distribution of the young lice, of kerosene oil emulsified with either
milk or soap, and properly diluted according to the directions given in
the reports of the Entomological Division of the Department of Agriculture at Washington, and in various other entomological publications.

The young lice were hatching from the eggs at the time that the specimens were sent.

AN UNRECOGNIZED APPLE ATTACK.

Under date of January 25, 1884, Secretary Harison informed me that he had been addressed by Mr. H. C. Watson, of Port Kent, N. Y., in relation to a great loss of apples in his vicinity recently. Mr. Watson had written as follows:

The apples affected, I think, exhibit no orifice, but appear on the surface round and fair. When cut across longitudinally, a circle of dark spots is revealed similar in position to those that children sometimes call "the ten commandments." Immediately upon exposure to the atmosphere a serum exudes from these spots, and generally a white worm or maggot soon appears in the serum.

Mr. Watson also stated to Mr. Harison that while almost all gardens in his village had the fruit affected in the manner above stated, yet some orchards had wholly escaped.

Under date of February 8th, Mr. Watson sent to me three of the apples showing the attack. Two of them had been cut transversely several days previously to all appearance, as the cut surfaces had partially dried, and showed the commencement of decomposition in the more moist portions. The "ten commandments" could be made out, but were not a conspicuous feature, and seemed to have no connection with the attack. Some small grains of larval excrement were scattered over the surface, but no larvæ were visible. A number of spots, small and blackish, could be seen.

Upon the fourteenth of February the cut apples were again examined, when, upon the surface of one, two larvæ of two species were discovered. The one, a white, slender larva, about one-tenth of an inch long, with its anterior end extended into an extensile point, evidently one of the Diptera, and perhaps of the genus *Ampelophila*. The egg from which it proceeded may have been deposited upon the fruit after the commencement of its decay.

The other larva was about one-twentieth of an inch long, with a large, flattened head, with strong jaws and conspicuous antennæ, with six legs as long as the width of the body, the body with a few short hairs, two longitudinal yellowish subdorsal stripes upon its posterior segments, and the last segment terminating in two short pointed processes. It evidently belonged to the Coleoptera. No explanation presents itself for its presence in the apple at this time.

Upon cutting the third apple, a number of spots of different sizes, from one-tenth of an inch downward, were observed, most of which were near the exterior. These spots contained yellowish, soft matter, and upon following some of them, they proved to be continued within the fruit in irregular channels. They may have been caused by the burrowing of small larvæ, but no evidences of this were discovered. After the above slight examination the two halves of the apple were placed together to await other examinations, which, however, were not made.

It is not probable that either of the above larvæ were the cause of the injury to the fruit of which inquiry was made.

REMEDIES AND PREVENTIVES.

COAL ASHES FOR THE CURRANT WORM.

From the New York Agricultural Experiment Station the following experiment of mulching currant bushes with coal-ashes is reported:

A plat of bushes mulched with this material in the spring, on which no insecticide application had been made, suffered less from the currant worm than an unmulched plat that had been several times treated with hellebore.

As the larva of the currant saw-fly transforms to the pupa but slightly buried beneath the surface of the ground, it would not be strange if the presence of the coal-ashes should prevent the development of the perfect insect, through either its chemical effect upon the delicate pupa or a mechanical action upon the larva in its burrowing into the soil.

A CARBOLIC WASH FOR THE PEACH-TREE BORER.

A wash prepared from the carbolic soap of Buchan & Co. (Messrs. Kidder & Laird, 83 John street, New York, agents), appears, from the testimony given in its favor, to be an effectual preventive of the attack of the peach-tree borer, *Ægeria exitiosa*. It is prepared as follows: A five-pound can of the soap known as the "Carbolic Plant Protector," to be ordered of the agents if not purchasable at the city drug stores, is to be emptied into a barrel (when a large number of trees are to be treated), upon which two or three pailfuls of hot soft water are to be poured. Let it stand for twelve hours or more to dissolve. When it is to be used, fill the barrel with cold soft water — making about thirty gallons of the wash, which will be sufficient for about three thousand trees of the ordinary size.

The following statement in regard to this wash has been made by Mr. M. B. Bateham, of Painesville, O., in the *Country Gentlemen* for August 24, 1876 (page 535):

Ten years ago, having thirty acres of young peach orchard, I found as others have done, that fighting the borers by semi-annual examinations of each tree, and digging out the worms with pen-knife or wire, was going to be a serious task. I then procured a lot of roofing-paper, of which coal-tar was an ingredient; cut it into strips six inches wide, and fastened these tightly around the base of each tree with a tack. This method was in the main successful, and much less expensive than the hand-killing. But a couple of years after, I received from a firm in New York a small box of "Buchan's Carbolic Soap," which was at that time extensively advertised as a remedy for plant-lice and other kinds of insects that annoy horticulturists, and I was requested to experiment with it according to hints and directions furnished.

I tried this soap on aphis, scale-lice and various other small insects, with good success, and was especially pleased *to find it a complete protection against the peach-borer*, which was of the most importance to me. The first season I applied it twice; first about the middle of June, as soon as I began to see any Ægeria moths flitting about, and again about a month later, as I had noticed that the insects continued to appear and lay their eggs for a month or longer time. But since the first season I have only used one application of the wash, about the first of July, each season, and not more than from three to five per cent of my trees have been affected with borers, and those not perceptibly injured by them. In a locality where peach orchards are plenty and the borers troublesome, I would recommend two applications of the wash each summer, and *I am confident it will prove completely effective* if the materials are good and the work properly done.

I have never been able to perceive the least injurious effects of this wash, even on young peach-trees, and I presume that its strength might be safely increased if thought desirable.

For aphis and bark-lice of the various kinds, I have found the carbolic wash of much service. It should not be used quite as strong when applied to the foliage of young shoots, as recommended for the borers. It is well to test the strength by wetting a few leaves with it, and noticing the next day whether they are seemingly injured by the application. If so, dilute it more before using.

I think, from repeated experiments, that the wash is as effective a preventive of the apple borers as the peach. For the old or eastern borer (*Saperda*), which works near the roots, the mode of application is the same as on peach-trees; but for the western, or flat-headed borer (*Chrysobothris*), the whole of the trunk of the tree, especially the south-west side and large limbs, if exposed to the hot sun, or where the bark is at all injured, should be washed.

Where the "Carbolic Plant Protector" cannot be conveniently obtained, a wash may be prepared with the carbolic acid to be purchased of druggists, which, according to Mr. Bateham, is equally effective with the manufactured carbolic soap.

4

In a gallon of common soft soap, thinned with a pailful of hot soft water, a half-pound of carbolic acid (the pure) is stirred.   After it has stood for a day or longer, or until the proper union has taken place, add thirty gallons of cold soft water, making a barrel of the wash. If a less amount of the wash is needed, of course the proportionate quantities of the materials may be used.   The crude carbolic acid would doubtless answer as well, but a larger quantity, perhaps double the amount, would be necessary.

Later, under date of April 15, 1880 (*Country Gentleman*, xlv, p. 246), the following somewhat stronger wash is given by Mr. Bateham, and it may be presumed to be that which continued experiments have shown to be the most effectual:

For an orchard of five hundred bearing trees, we buy a pint of crude carbolic acid, costing not over twenty-five cents (or half as much as the refined), then take a gallon of good soft soap and thin it with a gallon of hot water, stirring in the acid and letting it stand over the night or longer; then add eight gallons of cold soft water, and stir. We have then ten gallons of the liquid ready for use. Some peach growers use a little more and some a little less of the acid, but if it is much stronger than the above, it would be apt to injure the trees.   The wash should be thoroughly applied with a swab or brush around the base of each tree, taking pains to have it enter all crevices.

The proper time to apply the wash is about the last of June, if the weather is hot, or the first of July. Mr. Bateham, whose long experience in the cultivation of peaches has made him familiar with the parent moth and observant of its habits, has never seen one in his locality (Painesville, Ohio, N. latitude 41°, 40′) depositing its eggs before the first of July. He finds this to be the best time to apply the wash, "as it drives off the moth by its odor, and instantly kills any eggs that may have been deposited." For the apple-tree borers he applies this same wash about the first of June.

Pyrethrum for the Cabbage Worm.

At the New York Experiment Station, a mixture of one part powdered pyrethrum with three parts of plaster or air-slacked lime, has been found quite effective in destroying *Pieris rapæ* larvæ on cabbages. It is applied with a wooden bellows manufactured for the application of powdered insecticides, by inserting the nozzle among the leaves, so that the powder is driven through the plant.

Another mixture of pyrethrum, still further diluted than the above, and therefore more economical, is one part of the powder to twenty of flour, applied with a bellows. Experiments made with this preparation, show the *P. rapæ* larvæ to have been killed in twelve hours.

CARBOLIC ACID FOR THE CABBAGE WORM.

As cheaper than pyrethrum powder and more quickly applied, the following application may be made: One tablespoonful of the cheapest black carbolic acid diluted in one gallon of water, applied sparingly, after heavy rains, at intervals of three or four weeks, if the caterpillars are observed. Persons who have tested the above, claim that it has given them uninjured crops of cabbage. (*Country Gentleman.*)

ROAD DUST FOR THE CABBAGE WORM.

A simple and costless palliative, if not a preventive of the ravages of this insect, has been found in the frequent use of dry dust. A quantity of road dust should be kept in a covered vessel near the cabbage patch in the garden, and every few days or after rains, a little of it should be sifted over each cabbage head.

COAL-OIL REFUSE FOR THE CANKER WORM.

Instead of the tar or printers' ink band, applied about the trunks of trees to prevent the ascent of the wingless *Anisopteryx vernata*, the residuum from kerosene oil works may be used. It is cheaper and will last twice as long as the tar or ink, not requiring renewal oftener than once in six days. It has been used with success when printers' ink has failed. (*Transactions of the Massachusetts Horticultural Society* for 1883, Part 1, p. 16.)

A WASH FOR THE APPLE-TREE BORER.

The best protection from the injuries of the flat-headed apple-tree borer, *Chrysobothris femorata* Lec., so far as known at the present, is a wash made of soft soap and carbolic acid. Soft soap and lime, with a little glue dissolved to promote adherence, is also used. As this insect, unlike the round-headed borer, *Saperda candida* Fabr., formerly known as *Saperda bivittata* Say, does not confine its attack to the base of the tree, but extends into the branches, and even into the terminal twigs, as I have observed when the tree has become diseased, it is important that the soap be applied to such portions of the branches as can be reached. This coating, made in May, and again in early July and late August, will usually prevent the deposit of the eggs; but it is not well to trust alone to its efficacy. Where the insect abounds, the trees should be carefully examined for the presence of the borer, which may be detected by the excrementa or borings at the commencement of the burrow, or, when the borer has just entered, by an exuding drop of sap upon the bark at its place of entrance. A gentleman who has had much experience with this borer in the West, gives this advice (7th Rep. Ins. Mo., p. 79): "It is best for those having trees subject to attack, to look them over every week, if possible, or every two weeks at least, from the first of June until autumn, for exudation of sap from the bark, which is a sure indication of their presence. When noticed, the borer may be removed by cleanly cutting out a small slice of bark."

The above means involve labor, but if not resorted to, apple-trees cannot be grown in localities where this borer abounds, particularly in portions of several of our Western States, where as many as a hundred of the borers have been taken from one small tree.

SALTPETRE FOR THE STRIPED CUCUMBER BEETLE.

Saltpetre dissolved in water has been recommended for use against both the larva and the perfect insect of the striped cucumber beetle, *Diabrotica vittata* (Fabr.), and also for protection from cut-worms, as appears in the following paragraphs taken from agricultural journals.

To destroy bugs on squash and cucumber vines.—Dissolve a table-spoon-ful of saltpetre in a pailful of water, put a pint of this around each hill, shaping the earth so that it will not spread much, and the thing is done. The more saltpetre the better for vegetables, but the surer death to animal life. The bugs burrow in the earth at night but fail to rise in the morning. No danger of killing any vegetables with it: a concentrated solution applied to beans makes them grow wonderfully.

Wetting the soil around cucumbers, squashes and other *Cucurbitæ* three or four times at intervals of as many days, with a solution of saltpetre — an ounce per gallon of water — is referred to as preventive of ravages of the borer; and sprinkling the leaves with the same is suggested as useful against "the striped bug."

One year ago I had a patch of beans entirely destroyed by cut-worms. I planted it over; as soon as they came up the worms began again. I dissolved half a pound of saltpetre in three pints of water, mixed that thoroughly with one-half bushel of dry ashes, and sprinkled the ashes on the beans just as there was a shower coming up and the rain washed the ashes all off into the ground. I had no more trouble with the worms but had a good crop of beans.—[R. K., Franklin Co., Mass.

A saltpetre solution is frequently recommended to be poured about the roots of plants infested with cut-worms. It kills the pests and is a nitrogenous fertilizer.—(*New England Homestead.*)

The experiment with this material can so easily be made, that it deserves trial with several of our smaller root-insects, as for example, upon the radish, cabbage, and onion maggots, *Anthomyia raphani* Harris, *Anthomyia brassicæ* Bouché and *Phorbia ceparum* (Meigen).

Notwithstanding the statement of its efficacy against cut-worms above given, it may well be doubted whether the benefit supposed to have been derived from its use could not better be accounted for, from a cessation of the attack through the maturity and pupation of the larvæ after a second planting.

GYPSUM AND KEROSENE FOR THE SQUASH-BUG.

The following application for preventing the attack of the squash bug, *Anasa tristis* (De Geer), is given: To two quarts of gypsum put one tablespoonful of kerosene oil; this sprinkled on the vines will generally answer for the season. If the bugs return repeat the operation.

The person giving the above, states: I applied it this season on several thousand hills of melons, cucumbers, etc., after the bugs had commenced operations, and have not since had a vine destroyed. I have used it for several seasons with the same result. This is safer and cheaper than Paris green.

This preventive promises to be effective against the striped cucumber bettle (*Diabrotica vittata*), and the cucumber flea-beetle (*Crepidodera cucumeris*), which in addition to the squash, injure seriously the melon and the cucumber.

VAPORIZED TOBACCO JUICE FOR "THRIPS," ETC.

The vapor from tobacco juice for killing the smaller insects that are the pests of plant-houses, has lately been employed very successfully in France, but, we believe not to any extent in this country. If its merits are what are claimed for it, it will supply a long-needed want in our conservatories and graperies. One who has tested it states: "Ever since I adopted it, it has been absolutely impossible to find a thrips in my houses; and other insects have likewise disappeared." The method of using it is thus described:

Every week, whether there are insects or not, I have a number of braziers containing burning charcoal distributed through my houses. On each brazier is placed an old sauce-pan containing about a pint of tobacco juice of about the strength of 14°. This is quickly vaporized, and the atmosphere of the houses is saturated with the nicotine-laden vapor, which becomes condensed on every thing with which it comes in contact — leaves, bulbs, flowers, shelves, etc. When the contents of the sauce-pans are reduced to the consistency of a thick syrup, about a pint of water is added to each, and the vaporization goes on as before. I consider a pint of tobacco juice sufficient for a house of about 2,000 cubic feet. The smell is not so unpleasant as that from fumigation, and the tobacco juice can be used more conveniently than the leaves. Plants, no matter of what kind, do not suffer in the least, and the most delicate flowers are not in the slightest degree affected, but continue in bloom for their full period, without any alteration in their appearance. When the operation is completed, if the tongue is applied to a leaf, one can easily understand what has taken place from its very perceptible taste of tobacco.

The process requires to be repeated in proportion to the extent to which a house is infested. It is not to be imagined that these troublesome guests are to be quite got rid of by a single operation. A new brood may be hatched on the following day, or some may not have been reached on the first day, so that the vaporization should be frequently carried on until the insects have entirely disappeared, and after that it should be repeated every week in order to prevent a fresh invasion. (*Country Gentleman.*)

In France, the tobacco juice of the strength above stated, can be purchased of the tobacco factories at about twelve or fifteen cents (our money) a quart, by presenting a certificate that it is to be used for killing insects. The expense, at this rate, would be very trifling, being only about twenty-five cents a week for a plant-house fifty feet long by sixteen broad and ten high.

A strong infusion of tobacco leaves made by boiling, would be a substitute for the above factory juice. It might be prepared in quantity and evaporated for convenience of keeping and ready use, to the proper degree.

The " Thrips," for the killing of which the vaporized tobacco juice is recommended, is a popular name which has obtained currency among vine-growers for a small (about one-eighth of an inch long), slender, spindle-shaped, parti-colored leaf-hopper, which in its larval, pupal and perfect stages is very destructive to the foliage of grapevines. It attacks the leaves by puncturing them with its proboscis, usually upon the under side, withdrawing the sap and causing small discolored spots over their surface. The spots increase in size and number by coalescence and the growth of the insect, becoming later, large, brown blotches, which gradually extend, if the attack is severe, over the entire leaf, until it dries, appearing as if scorched by fire, dies, and falls from the vine. The fruit is dwarfed, fails to ripen, or the vine is killed, according to the abundance of the insects.

It is believed that several distinct species are usually associated in this attack, belonging to the genus *Erythroneura*, of which the principal one is that described by Dr. Harris in the year 1831, under the name of *Tettigonia vitis*, a figure of which (pronounced a poor one by Mr. Walsh*) is given in Plate 3, of *Insects Injurious to Vegetation.*

*A better one may be found in the *Practical Entomologist*, ii, p. 51, which has been copied in Packard's *Guide to the Study of Insects*, in Saunder's *Insects Injurious to Fruits*, and in other recent publications.

The Thrips proper is quite a different insect. There are a number of species, constituting the family of *Thripidæ*, the location of which, in classification, has caused much discussion. By Haliday, it was separated as a distinct order under the name of THYSANOPTERA, which has been accepted by a number of entomologists. Dr. Packard and others regard it as belonging to the HEMIPTERA, having affinities both with the *Corisidæ* and the *Mallophaga*. Their habits vary greatly, for while many of the species are certainly vegetable feeders and injurious in their operations, others are carnivorous, and are serviceable in the destruction of gall-insects,[*] eggs of the curculio,[†] the red-spider (*Tetranychus telarius*).[‡] the clover-seed midge (*Cecidomyia leguminicola*)[§] probably the wheat-midge (*Diplosis tritici*),[‖] and other insect pests.

INFUSION OF TOBACCO FOR THE ROSE-LEAF HOPPER.

An infusion of tobacco is said to be an effectual preventive of the little white leaf-hopper often so injurious to roses, the *Tettigonia rosæ* of Dr. Harris. Where the tobacco stems can be procured, place some of them in a vessel (a tin pan of the capacity of about two gallons would be convenient) and pour boiling water over them, so as to cover completely, and leave it standing over night. Dilute for using with four or five times the quantity of water and apply with a syringe or force pump, taking care to distribute it also over the underside of the leaves. The application, like most other liquid applications to leaves for the prevention of insect injuries, should be made in the evening or early in the morning. It should first be applied early in the season before the injuries are very apparent, and as soon as the young larvæ, looking like little white specks, can be discovered upon the underside of the leaves. As often as may be needed, in order to check the attack, the showering with the infusion should be repeated.

INFUSION OF TOBACCO FOR APHIDES ON HOUSE-PLANTS.

A similar infusion to the above is recommended by Mr. John G. Barker, at a recent meeting of the Massachusetts State Horticultural Society, for killing the "green fly" (Aphis) that infests house-plants. It is made by filling a pail with stems and pouring on them all the water the pail will hold. This should stand for twenty-four hours and used in the proportion of half a pint to a pail of water [seemingly a very

---

* Riley: 6th *Report Insects of Missouri*, p. 50, and 5th do., p. 119.
† Id.: 2d Report, p. 6; 3d Report, p. 29 (the species not named).
‡ Pergande, in *Psyche*, iii, 1882, p. 381.
§ Id., ib.
‖ Walsh, in *Proceed. Entomolog. Soc. Phila.*, iii, 1864, p. 611.

weak infusion]. The plants should be turned bottom up, placing the left hand over the top of the pot to prevent accident, and then plunging it in the infusion once or twice until the insects drop off. Some of the liquid should always be kept on hand and used on the first appearance of the aphis. After using, the plant must be rinsed in clean water of the same temperature as the room. A florist in Philadelphia kept his plants free from injury in this way without fumigating.

KEROSENE OIL FOR SCALE-INSECTS ON HOUSE-PLANTS.

Mr. Barker had found the scale-insects so common on oleanders (probably *Aspidiotus nerii* and *Lecanium hesperidum*) and other thick-leaved plants, more difficult to be destroyed. These may be washed with whale-oil soap and water. A sponge dipped in a little sweet or kerosene oil, and wiped up the stem and under the leaves occasionally, will keep off the scale effectually. This had been applied to plants which had been neglected and became very dirty, using the kerosene so freely that there were misgivings of the result, but with only beneficial effects.

SULPHUR FOR CABBAGE APHIS.

A correspondent of the *Home Farm*, states that after having contended for several years with cabbage-worms and lice, and using nearly all the various remedies proposed for them, he has at last succeeded in ridding his cabbages from attack, by a means which he thinks is new, and which he hastens to give to the public. It is simply to sprinkle a little sulphur on the heads of the cabbage when the presence of the insect is noticed. He had applied it three times during the season and " deliverance came every time, at once." The sulphur caused the leaves to curl slightly and to discolor a little, where it was distributed the thickest, but it did not check the growth nor otherwise injure the plants.

GRAPE BAGGING TO PROTECT FROM INSECT ATTACK.

The utility of bagging grapes, is a question not yet settled by grape-growers. Perhaps no general rule can be established, under the greatly varying conditions of soil, culture and exposure of different localities. If the fruit is liable to insect injury, it offers a means of entire protection, at a cost, it is believed, more than repaid, by the attendant benefits of protection from rot, finer color, richer bloom, larger size, etc.

The principal insect attacks which it would prevent are the following named: The grape-seed midge, *Isosoma vitis* Saunders — a small hymenopterous insect which lays its eggs upon the grape during the month of July, the larva from which burrows into the fruit to feed upon the seed, and causes the grape to shrivel and dry. The grape-berry moth, *Eudemis botrana* (Schiff.), which deposits its eggs in June, the larvæ from which destroy the grapes in July by feeding within them and binding them together with patches of their webs and excrementa. The grape curculio, *Craponius inæqualis* (Say) —a small snout-beetle which punctures the fruit for the deposit of its eggs, causing its premature ripening and dropping to the ground. Several species of the larger Hymenoptera, as the honey-bee, wasps and hornets, feed upon the ripening fruit.

The method of bagging, and the cost of the bags is told by a correspondent of the New York *Tribune:*

Our practice has been to bag very early after the blossom falls, which protects the fruit from all destroying depredators and influences, one only excepted, which is cracking when rain-storms last several days or a week; but such storms seldom occur when grapes are ripening. The bag is torn down just far enough on either side to lap the two ends over the branch above the cluster and a pin put through on the under side of the limb at each corner, being careful to draw the bag over the torn ends to exclude water. The lowest corner should be pierced with the small blade of a pocket-knife for drainage. This is better than a round hole made with an awl, as it closes and excludes the spores of fungi and small insects. We order two-pound bags for two-thirds of the number needed, for single clusters; for the other third, four-pound bags to be used over two clusters occurring together. They can be ordered from any paper warehouse as cheaply as from the factory. Two-pound Manilla bags cost $1.80 per 1,000; two-pound imitation Manilla bags, $1.50 per 1,000; and 10 to 15 per cent advance on each of the larger sizes. By counting the clusters intended to be bagged on an average vine, the approximate number needed can be readily ascertained.

A BOTTLE TRAP FOR VARIOUS GARDEN INSECTS.

Charles Downing mentions the following as an effectual trap for all sorts of garden insects. Fill wide-mouthed bottles half full of a mixture of water, vinegar and molasses, and suspend them among the trees. In a short time they will be full of insects, and must then be emptied and the liquid renewed. An acquaintance of his captured in this way, more than three bushels of insects in his garden in a season, and preserved it almost entirely free from their ravages.—J. W. Manning, in *Trans. Mass. Horticultural Soc.* for 1883, p. 14.

POULTRY IN ORCHARDS.

The presence of large numbers of chickens, turkeys and ducks in orchards and vineyards seems to be the most effectual means yet dis-

5

covered for combating the number of insect pests which are taxing the ingenuity and perseverance of orchardists and vineyardists. While these fowls will not undoubtedly keep vines and trees entirely free from pests, they render a vast amount of assistance in that direction. They are proving especially valuable in case of invasion of grasshoppers. Indeed, where this pest has not appeared in overwhelming numbers, fowls are the most effective remedy yet discovered for their destruction. Thousands of dollars have been saved to the fruit and vine-growers of this county the present season by the keeping of fowls in their orchards and vineyards. * * * * Ducks have not been used here much for this purpose, but are said to do very satisfactory work. They have the most insatiable appetites for grasshoppers of any of the domestic fowls. The raising of ducks in itself is very profitable. Fruit culture and poultry raising seem destined to become inseparable industries in this country. (*Pacific Rural Press.*)

INSECTICIDAL PROPERTIES OF SOME OF THE COMPOSITÆ.

Prof. F. G. Sanborn has written me as follows:

" I will here call your attention to the fact that several other of the *Compositæ* besides those that you have mentioned, notably, *Leucanthemum vulgare* (the ox-eye daisy) and *Maruta cotula* (the common Mayweed) have proved very useful insecticides and were continually recommended by me when acting entomologist to the Massachusetts State Board of Agriculture from 1858 to 1869, inclusive. They have been used in powder, also in tea made from the flowers without pulverizing, and with excellent effect. A strong tea made from the grocer's 'Cayenne pepper' has been applied to plants infested with mandibulate insects [furnished with mandibles for biting], with good results, and has also been used as a protection from carpet eating insects."

A NEW INSTRUMENT FOR USE AGAINST ROOT-INSECTS.

An instrument has lately been devised by Dr. W. S. Barnard by which those insecticides which are dangerous to plants, such as kerosene, cyanide of potassium [KCN], and bisulphide of carbon [CS₂], might be used with safety to the plants and the destruction of the insects in the ground. These substances have usually been applied on or just beneath the surface of the ground, either among or above the roots, but killing them when used in strength when coming in contact with them. When applied in volatile form they are less injurious. The instrument for conveying the insecticide beneath the roots, called a "nether-inserter," consists of a tube fitting closely around a solid shaft somewhat longer than the tube and pointed at its lower end, the tube being 15 mm. in diameter [0.6 inch] and the shaft 12 mm. [0.48 inch]. The upper end of the tube expands like a bowl. The upper portion of the shaft is weighted with a heavy ball, so disposed that the shaft can be grasped above the ball. By partly withdrawing the shaft from the tube and then returning it with force as the lower end of the tube rests on the ground, both can be driven into the ground to any required depth. The shaft is then wholly withdrawn, and the insecticide poured into the tube, thus placing it beneath the roots without coming in contact with them, if the instrument has been properly inserted. The tube is then withdrawn, and the hole made by it filled with earth. The insecticide, being volatile, rises through the ground and becomes diffused. With this method of application it is believed that kerosene is superior to napthaline. (*Psyche*, for January–February, 1884, iv. p. 134.)

In a communication by Dr. Barnard to *Psyche* for March, 1884, it is stated that experiments made with the nether-inserter in applying petroleum for destroying phylloxera on vines near Washington have given "perfect satisfaction." "The nether-upward kerosene diffusion process is the only economically practical way in which the deep application of the undiluted forms of petroleum can be attempted with safety to the plant. By it the cheap, crude article and its lighter form, the naphthas, become most valuable agents against the pests. It applies likewise as a treatment against all other root-insects or subterranean pests, as, for example, the American blight aphid [*Schizoneura lanigera*], the hop-root Gortyna [*Gortyna immanis*], root maggots of the cabbage, etc. [*Anthomyiidæ*], the strawberry root beetles, cicadas, cutworms, white-grubs, wire-worms, nests of ants, etc. Thus it is seen to have a general application to a wide range of cases heretofore not satisfactorily treated."

Dr. Barnard claims also that the nether-inserter may also be efficiently used with many other insecticides, of which are the following: "Rhigolene, gasolene, naphtha, benzine, kerosene, crude petroleum, oil of tar, tar water, naphthaline, pyroligneous acid, soot, creosote, carbolic acid, cresylic acid, sulphurous acid, sulphocyanide of potassium, bisulphide of carbon, cyanide of potassium, pyrethrum preparations, lye solutions, tobacco decoction, chips and snuff water, gas water," and it may also be used for the application of liquid fertilizers, vapors, gases or fumes. In short it is available with any upward-acting insecticide against any underground enemies.

NAPHTHALINE AS AN INSECTICIDE.

From a paper read by Prof. C. V. Riley before the Biological Society of Washington, on December 14, 1883, it appears that naphthaline, [$C_{10} H_8$] was first made in 1808. Its use as a substitute for camphor

36   SECOND REPORT OF THE STATE ENTOMOLOGIST.

for killing museum pests, was suggested in 1840. Placed in insect boxes, it kills acari and psoci, but not other museum pests. Experiments were made with it against *Phylloxera vitifoliæ* in 1872. Fischer began experimenting with it in 1881. It is a better insecticide and cheaper in its crude form than when pure, but is more injurious to plants in that form. It has been applied to grapevines by pouring a kilogram [2.20 pounds] of it in a trench from 15 to 20 cm. [6 to 8 inches] deep near the stock of the vine and then filling the trench with earth.

At a subsequent meeting (28th of December), Dr. T. Taylor stated that rats, mice, crickets and locusts were driven away by the use of naphthaline. Earth worms were driven out of the ground and killed by placing it in the bottom of a flower-pot where they occurred. Insects infesting seeds were killed by placing it in jars with the seeds, without injury to the seeds. Moistening increases the efficacy of naphthaline (*Psyche*, for January–February, 1884, iv, pp. 133–134).

KEROSENE OIL EMULSIONS.

In the preceding report of the Entomologist, reference was made (page 44) to an emulsion of kerosene and milk, but no directions were given for making it. The omission is here supplied. The method recommended by Mr. H. G. Hubbard, a special agent of the U. S. Department of Agriculture, seems to be the best presented. It is as follows:

Take of refined kerosene two parts, fresh, or preferably sour cow's milk, one part (percentage of oil 66⅔). Mix in a pail or tub, by continuous pumping with a force-pump back into the same vessel through the flexible hose and spray nozzle. After passing once or twice through the pump the liquids unite and form a creamy emulsion, in which finely divided particles of oil can be plainly detected. Continue the pumping until the liquid curdles into a white and glistening butter, perfectly homogeneous in texture and stable. The time required for producing the butter varies with the temperature. At 60° Fahr. it will be from one-half to three-quarters of an hour; at 75°, fifteen minutes; and the process may be still more facilitated by heating the milk up to, but not past, the boiling point.

Upon standing for a day or two the milk (if sweet, has been used) will curdle, but as it simply thickens and hardens without separating from the oil, it needs only to be stirred, not churned again, to bring it back to its former smoothness. If sour milk is used no fermentation ensues, and if not exposed to the air, the butter can be kept unchanged for any length of time.

When needed for use the butter will mix readily with any proportion

of water, if first thinned with a small quantity of the liquid. In using
the emulsion for killing scale-insects (perhaps the most difficult of insects to be destroyed by it), the kerosene butter should be diluted with
water from twelve to sixteen times, or one pint of butter to one gallon
and a half of water. Dilute only as needed for immediate use. (*Report of the Commissioner of Agriculture for the Years* 1881, 1882, pp.
113–114.)

A soap emulsion.—Dr. J. C. Neal, also a special agent of the U. S.
Department of Agriculture — as the result of his experiments with
various emulsions of kerosene oil, recommends very highly the following:

Four pounds of rosin soap — common bar or yellow soap, dissolved
in one gallon of water, with heat. Add gradually one gallon of kerosene
with constant agitation. A gelatinous compound is formed which is
very stable.

A gallon of this emulsion, containing fifty per cent of kerosene, costs
twenty-six cents.

Reducing the above by the addition of forty-nine gallons of water,
gives one per cent of kerosene, costing a little more than half a cent
per gallon.

This reduced emulsion is of a milky color and is quite permanent.
Dr. Neal was satisfied that with it, thoroughly applied, *all* the cotton-
worms reached by it, were killed without injury to the plants. (*Bulletin
No.* 1. *U. S. Dept. Agricul.*— *Division of Entomology,* 1883, pp. 32, 42.)

Later, Mr. H. G. Hubbard reports that further experiments made by
him with kerosene emulsions prove that various soaps can be readily
made to combine with the oil, and that the soap and kerosene emulsions
are as effective as those formed with milk. The use of soap materially
reduces the cost, except where milk is abundant and cheap.

Common bar soap, soft soap and whale oil soap have been tried and
found to be almost equally good. The following formula is one which
has proved in practice useful where a moderate quantity of emulsion is
required:

Kerosene.................................2 gallons = 67 per cent.
Common soap or whale oil soap ....½ pound, } 33 per cent.
Water....................................1 gallon,   }

Heat the solution of soap and add it boiling hot to the kerosene.
Churn the mixture by means of force-pump and spray nozzle for five or
ten minutes. The emulsion, if perfect, forms a cream, which thickens
on cooling, and should adhere without oiliness to the surface of glass.
Dilute before using, one part of the emulsion with nine parts of cold
water.

The above formula gives three gallons of the emulsion, and makes, when diluted, thirty gallons of wash. [*Report of the Entomologist : in Report of the Commissioner of Agriculture for* 1883, p. 152.]

The cost of the above wash, with kerosene oil at 12 cents the gallon, and yellow soap at 6 cents the pound, would be less than one cent per gallon.

The Shearer soap emulsion.—In a recent report, Miss Ormerod, Consulting Entomologist of the Royal Agricultural Society of England, presents the following recipe "for a simple and effective method of making a mineral oil solution," which was devised by Mr. Alex. Shearer, " a clever chemist as well as an able and intelligent gardener," and which, after experimenting with the emulsions as prepared in this country and elsewhere, she pronounces the best that she had met with. It has now been tried for several years, and found both safe and serviceable.

To eight parts of soft water add one part of black (soft) soap, and boil briskly for a few minutes until the soap is thoroughly dissolved. While boiling add paraffin [kerosene] or any other similar oil, and boil for a minute or two longer when the whole will be thoroughly amalgamated, and, if bottled and securely corked while warm, it will remain so and be fit for use at any time when required. For field use, the immediate application would save all need of storing. The strength of the solution depends on the amount of mineral oil in it, and it can be easily reduced to the proper power by mixing it with soft water as it is wanted for use.

The following notes are also given of the method found the most convenient for mixing the application.

Eight parts of water and one part of soft soap thoroughly amalgamated forms the lye which *takes* mineral oil, and thoroughly mixes with whatever proportion of the oil be added. As heat aids much in quickly producing thorough amalgamation of the ingredients, boil the soap and water together, and when ready, turn it into ordinary wine bottles (costing little or nothing) which have been placed in boiling water. About half fill the bottles, turn two gills of the oil in each bottle, then fill up with the boiling lye, cork at once, and store away for use.

When required for use, a bottle of the mixture is poured into a fourgallon watering pot which is filled up with soft water, and is ready for use, at a strength of one wine-glass of oil (half a gill) to one gallon of water [one part of oil to 64 of water — about $1\frac{2}{3}$ per cent of oil].

Half a gill of oil to a gallon of water is strong enough to kill Aphides and such soft insects ; one gill for " thrips " [the small leaf-hoppers of the grapevine probably], and a gill and a half for scale-insects.

By bottling the mixture as above, no mistake need be made in using it of the proper strength.

MISCELLANEOUS NOTES.

PARASITE OF PYRAMEIS ATALANTA.

A small parasite upon this butterfly, received from Mr. W. H. Edwards, of Coalburgh, Virginia, was identified by me as *Microgaster carinata* Packard, described in the *Proceedings Boston Soc. Nat. Hist.*, xxi, 1880, p. 25–6. Upon submitting it to Prof. Riley, it was compared with the type of *M. carinata* in his possession and found to be identical. It should, however, he stated, be regarded as a variety of his *Microgaster gelechiæ* (First Missouri Report, 1869, p. 178) obtained by him from the gall-making caterpillar of the solidago gall-moth, *Gelechia gallæsolidaginis*, from which " it differs only in the black anterior and intermediate coxæ and trochanters, and darker tarsi and tips of tibiæ — all variable characters within the same species."

The *P. Atalanta* parasite would therefore be — conforming to a later generic arrangement,— *Apanteles gelechiæ* Riley.

Prof. Packard has also described (*loc. cit.* p. 27) a second parasite from the same butterfly, as *Microgaster Atalantæ.*

A DISEASED BROOD OF ACTIAS LUNA.

One hundred and seven eggs were deposited by a female brought to me, on June 27–8. They hatched July 9th, and all molted for the first time on July 15th. Five examples, just commencing to molt — the head withdrawn from the head-case — were placed in alcohol.

The 2d molt commenced on the 18th, with 5 individuals; on the 19th 19 molted, of which 4 were put in alcohol and 2 died; on the 20th 37 molted, and 1 died; on the 21st 7 molted and 2 died; on the 22d 2 molted and 2 died. Seventy passed their molt and seven died.

The 3d molting commenced July 23 — 3 individuals; on the 24th, 12; on the 25th, 24, and 7 had died.

After this time, the remainder were so diseased and died in such numbers, that no further record was kept. They were supplied frequently with fresh hickory leaves for food, but without avail. Several passed the 4th (the final molt), but died one after another — the last when apparently about three-fourths grown.

The indications of approaching death would be, eating sparingly, assuming a duller shade of green, and changing to a brownish color. They would be found dead the following day, black, perhaps hanging limp from a leaf or stem, and drawn out in length, the entire contents of their body semi-fluid, so that it was difficult to remove some of the examples without breaking them and discharging the black fluid.

The disease was probably identical with that which had frequently been observed by me in attempting to rear larvæ of other species of Lepidoptera in confinement, and by other entomologists, and with that which was studied by Professor Forbes, in 1883, when a very large percentage of the caterpillars throughout the northern and eastern portions of Illinois were killed. It was found by him to be a contagious disease, resulting from the presence within the larvæ of innumerable bacteria, probably *Micrococcus bombycis*, and in no ways distinguishable from the *flachéri*, which was investigated by Pasteur with such brilliant results at the time when, as a silk-worm disease, it was inflicting an annual loss of hundreds of millions of dollars annually upon the silk industry of Europe. See Prof. Forbes' address " On a Contagious Disease of Caterpillars," delivered before the State Horticultural Society of Illinois, in December of 1883.

The same disease has also been noticed by me, as destroying *Pieris rapæ* larvæ in cabbage fields in the vicinity of Albany.

HETEROPACHA RILEYANA *Harvey*.

Mr. G. R. Pilate, of Dayton, Ohio, sent under date of July 22d, some small larvæ of this species and also some eggs, found by him a few days previous. Not being able to rear them, in subsequently communicating with him, I learned that he had found no difficulty in breeding the species by the following method: "I use large glass candy-jars, eighteen inches high by eight wide, for my larvæ, putting in them about an inch and a half of white sand, slightly moistened, for if too wet it will kill the larvæ. For those that bury, I use three inches of sand. Upon the sand a piece of paper is placed to hold the excrement and permit of its easy removal when fresh food is supplied. Always give them Honey-locust to eat (*Gleditschia triacanthus*) and not the black locust."

Later, additional eggs were sent which had been laid the 8th of August, with the statement that "when first laid they are pure white, with the exception of the three spots on the top and sides."

HEMILEUCA MAIA (*Drury*).

Mr. Pilate informs me that from about three hundred (eggs?) of this species, found on basket-willow (*Salix viminalis*), he had obtained about one hundred pupæ. The species appears to be an easy one to rear in confinement, for in colonies reared by me from the egg-belts, nearly all attained their perfect stage. Larvæ collected at somewhat an advanced age upon their food-plants are quite liable to have been parasitized, for of a cluster of thirty found by me after their second molt, about one-third were subsequently destroyed by ichneumons, which proved to be *Limneria fugitiva* (Say) and an undetermined species of *Microgaster*. For details of this parasitism, see my " Biography of Hemileuca Maia " in the *Twenty-third Report on the N. Y. State Cabinet of Natural History*. 1872, pp. 146, 147, or *Entomological Contributions* [No. 1], pp. 14, 15.

THE HOP GRUB — GORTYNA IMMANIS (*Guen.*).

In my First Report, the reference to the " grub " attacking the root of the hop-vine (page 61), has been interpreted by an English writer as referring to the "white-grub," *Lachnosterna fusca* Frohl. — a very natural interpretation, from the well-known habits of that destructive species as a root-devourer, taken in connection with the customary limitation by our more careful writers of the term "grub " to a coleopterous larva, and often to the conspicuous and well-characterized larva of the *Scarabeidæ*.

The "hop-grub " was long supposed to be the larva of a beetle allied to the "white-grub," if not that identical species. Even so late as May of 1882, a paper was published in the Canadian Entomologist,* by Mr. Charles R. Dodge, of Washington, D. C., upon the "Hop-vine Borer," in which the feeding habits of the "larva " are correctly given, together with much valuable information of the injuries committed by it, and best method of dealing with it, but without presuming so much as to venture an opinion even as to the Order to which the insect might belong. Our first knowledge of its true character was that obtained from Professor J. H. Comstock, who, at the Annual meeting of the Entomological Society of Ontario, held at Montreal, during the meeting of the American Association for the Advancement of Science, in August of 1882, exhibited to those in attendance several examples of the insect which he had succeeded in rearing from the "hop-grub." It was a large and conspicuous moth, evidently belonging to the genus *Gortyna*, and presumably referable to the *Gortyna immanis* of Guenée.

We have no knowledge of any publication by Professor Comstock of his study of this insect. It has subsequently been carefully studied by Mr. John B. Smith, of Brooklyn, L. I., under the direction of the Entomological Division of the U. S. Department of Agriculture, who has given its life-history, together with remedies for its ravages, in *Bulletin No. 4, Division of Entomology*, 1884, pp. 34–39, figs. 2–4. The following is a summary of its history:

The egg, yellow-green, round, and of the size of a pin-head, is depos-

---

Canadian Entomologist, 1882, xiv, pp. 93-96.

6

ited upon the hop-vine as it begins to climb. It hatches in a few days
and the slender, black-spotted, greenish larva produced, burrows into
the vine just below the tip, soon causing the vine to cease climbing,
point downward, and almost to stop growing. When the larva has
attained a length of about half an inch, it emerges from the tip, drops
to the ground, and enters the stem at the surface of the vine, where it
feeds upward for a time. When it has grown to about an inch in length,
it changes its direction and burrows downward to the base of the vine,
at its junction with the old stock, and, eating its way out, completes its
growth as a subterranean worker. The journey from the stem to the
ground is made in the beginning of June, and before the 20th of the
month. Here it eats a small hole into the side of the stem just below
the surface and immediately above the old root. The hole is gradually
enlarged until the vine is barely attached to the root, or, as sometimes,
entirely severed from it.

By the middle or the 20th of July, the larva (two inches in length)
has matured, when it transforms to the pupa state in a rude cell, close
to the roots of the plant.

Mr. Smith states that the insect hibernates as a pupa, although "a few
specimens of the moth appear in the autumn, but the majority appear in
the spring, from the beginning to the end of May or later, according to
the season. Whether the former hibernate or whether they perish, I
have not been able to ascertain, though the latter seems the more
likely."

My own observations, and such records of the moth as are accessible
to me, indicate its late summer appearance only, and consequently, by
inference, hibernation in that stage. My collections of it have been
made only between August 25th and September 6th; and the examples
in the collection of Mr. W. W. Hill, of Albany, bear the date of
August 15 and 26. It is recorded as having been taken by Mr. O. S.
Westcott, at Maywood, Ill., on August 26, September 3, 4, 10, 11, 22, 23
(*Canadian Entomologist*, viii, 1876, p. 15); by Mr. Roland Thaxter, at
Newton, Mass., in August (*Psyche*, ii, 1877, p. 36); and by Mr. C. E.
Worthington, at Chicago, Ill., also in August (*Canadian Entomologist*,
xi, 1879, p. 69). The species does not appear in the published lists of
collections of *Noctuidæ* of Messrs. Norman (at St. Catharine's and Orilla,
Canada), Hill (in the Adirondack Region of New York), Devereaux
(at Clyde in Western New York), Pilate (at Dayton, O.), and Prof. Snow
(of Eastern Kansas).

THE "ARMY-WORM" IN WESTERN NEW YORK.

An appearance of the army-worm, in the vicinity of Clyde, N. Y., during the month of July, in 1883, was reported in several newspapers. Upon requesting of Mr. W. C. Devereaux that he would endeavor to ascertain what the insect was, as there was no probability of its being the true army-worm, *Leucania unipuncta* Haworth, he replied:

The worm, as you supposed, is not the *Leucania unipuncta*. Its feeding was done entirely on the foliage of swamp trees, taking a broad belt, and working on the very highest ones, consuming the entire foliage except the midribs. It showed a preference for ash and soft maple, although it fed on elms coming within its range. It has been said that it was a " measuring-worm," but no webs or tents were seen.

Mr. Devereaux was prevented from visiting the locality where the insect occurred, and was therefore unable to give any information of ·its true character.

The moth of the army-worm is a rather common insect in the State of New York, and is known to collectors as an annoying visitor in their " sugaring " operations; but it is seldom that the larva presents itself in injurious numbers, so as to merit the appellation of the " army-worm." It seems, however, to have made a formidable demonstration, in August of 1882, in the neighborhood of Saratoga Springs. Examples of a black caterpillar (about one inch long, with two stripes the length of its body) which were represented as having destroyed twenty-five acres of meadow in the town of Saratoga Springs, were sent by Mr. F. D. Curtis, of Charlton, to the Department of Agriculture at Washington, for name. Answer was given, that "although badly shriveled and almost unrecognizable, they seem without doubt to be the genuine army-worm " (*Bulletin* No. 2 — *Division of Entomology* — *U. S. Dept. of Agriculture*, 1883, p. 28).

As injuries from this insect have been recorded in but few portions of the State of New York, it is of interest to note in connection with the above statement that Saratoga county is among the few places where it is believed to have previously occurred. Dr. Fitch, in mentioning its appearance, as the *black worm*, in Worcester, Mass., in May of 1817, adds, quoting from the *Albany Argus :*

This black worm is also destroying the vegetation in the northern towns of Rensselaer and eastern section of Saratoga. Many meadows and pastures have been rendered by their depredations as barren as a heath (6th–9th *Reports Ins. N. Y.*, 1865, p. 116).

The only other New York localities of its occurrence given by Dr. Fitch in his excellent account of this insect, *loc. cit.*, pp. 113–126, are the following:

Here in our own State, the worm has appeared in the vicinity of Buffalo, and at several other points toward the western and southern line of the State; and also in numerous places on Long Island. The State Agricultural Society has received * * * specimens of the worms from the town of Dix, near the head of Seneca Lake, where they were discovered August 12th, and of cornstalks and grass as ate by them (*ibid.*, p. 117).

ANISOPTERYX VERNATA (*Peck*).

Messrs. L. & A. B. Rathbone, extensive fruit-growers, at Oakland, Genesee Co., N. Y., write as follows:

" From 1873 to 1878, the canker-worms made sad havoc in our large apple orchard of a thousand trees. After spending several hundreds of dollars in trying to destroy them, we sprayed them with London purple, and have never had to do it since."

The above appears to have been a local attack of the insect, and its arrest by the means described shows how easily it may be accomplished and its further spread prevented. While yet extremely local within our State and not of frequent occurrence, it is of the utmost importance to fruit-growers that it be not permitted to increase and extend until it shall become established in our elms, as in New England, where, from the size of the trees, it will be almost beyond control.

THE CARPET-FLY.

From Mr. C. A. Richardson, of Canandaigua, N. Y., were received under date of Jan. 5, 1883, specimens of "a very destructive moth found in large numbers under a carpet." They were the vacated cases of *Tinea pellionella* Linn. — the only case-bearing Tineid feeding on woolen fabrics that occurs in the United States. In the same box were contained several long, slender, white, worm-like forms, which were in all probability the larvæ of the carpet-fly, *Scenopinus fenestralis* (Linn.), subsequently re-described as *Scenopinus pallipes* by Say. Under the latter name it has been described and figured in the *Guide to the Study of Insects*, 1869, p. 401, f. 322, in both the larva and the imago stages. Dr. Packard gives the following account of it:

The larva is found under carpets, and is remarkable for the double segmented appearance of all the abdominal segments, except the last one, so that the body, exclusive of the head, seems as if twenty-jointed instead of having but twelve joints. The head is conical, one-third longer than broad, and of a reddish-brown color, while the body is white. It is 0.65 of an inch in length. The larva is also said to live in rotten wood, and is too scarce to be destructive to carpets. The fly is black with a metallic hue and with pale feet.

Mr. Richardson, in his communication, suggests that the larva may perhaps feed upon the larvæ and the pupæ of the carpet moth, with

which he found it associated.   The doubt that exists in regard to its
food, shows the desirability of further study of the insect.

Its identity with the European species has been shown by Dr. Loew,
in Silliman's Journal, N. S., vol. 37, p. 318. Mr. Glover, in his MS.
Notes on the Diptera, presents figures of it, and remarks that the fly is
" very common in windows in the spring."

The structural characters of this insect have made it impossible to
give it proper classification.   From the peculiarities that it shows it has
been taken by Dr. Loew as the type of a separate family, viz., *Scenopinidæ*.   Of this family, its affinities are at present undetermined, but it is
thought that it may be related to *Bombylidæ* (*Diptera of North America*,
Pt. I, 1862, p. 28).   Four United States species only have been included
in the *Scenopinidæ* by Baron Osten Sacken, examples of which are contained in the Museum of Comparative Zoölogy, at Cambridge, Mass.
(*Catalogue of Diptera of North America*, 1878, p. 97).

THE EMASCULATING BOT-FLY.

A larva taken from the scrotum of a striped squirrel, *Tamias striatus*,
and agreeing in all particulars to the description given by Dr. Fitch
of *Cuterebra emasculator*, in his Third Report (page 162 of Reports
iii–v), was brought to me on the 13th of September, 1858, at Schoharie,
N. Y. A second one had been left in the squirrel, an end of which
could be seen, it was stated, through a small opening in the scrotum,
made undoubtedly for the purpose of respiration.

The example brought me was placed upon some ground mixed with
sand to give it less compactness, and although seemingly quite weak
from having been kept for two days wrapped in paper, it buried itself
out of sight in the course of two hours.

Two days later the squirrel was found where it had been left, with
the grub within it and alive, lying opposite to the place from which
the other had been taken, and with the tip of its body showing through
the opening.   Upon enlarging the opening with the point of a knife
and pressing slightly, the larva slowly emerged.   It was of a smaller size
than the first, being but about three-fourths of an inch in length.   The
testicle in which it was located, had been entirely consumed.   It was
given some loose ground in which it buried, but reappeared upon the
surface two or three times before its final burial.

Both of the above were probably immature when obtained, for neither of them developed the imago.

An alcoholic specimen of the larva, also taken from the scrotum of a
striped squirrel, is in the New York State collection.   It measures 0.75
inch long by 0.36 inch broad, and shows distinctly the peculiar granulated
surface, the rounded ten segments, and circular mouth-parts.

For an interesting account of this insect — one of our twenty-three known species of *Œstridæ*, see the report of Dr. Fitch above cited, pages 160–167. An account of the occurrence of the larvæ (three individuals) of an allied species, *Cuterebra buccata* (Fabr.), within the body of a striped squirrel, in the region of the kidneys, is given in the *American Entomologist*, i, 1869, p. 116–17, by Mr. S. S. Rathvon. The opinion is expressed by the editors that the larvæ may have emasculated the squirrel before their transferral to another portion of the body for food.

BEET-LEAF MINING ANTHOMYIIDÆ.

W. S. Miller, M. D., of South Britain, Conn., writes under date of June 9, 1884, that he is informed by an observing farmer that his beet-leaves had been infested by these larvæ for the last six or eight years [noticed first in New York in 1881], and that they had prevailed to an extent to cause the gardener much trouble. They were just commencing for the present year. A similar larva had been found by Dr. Mille in the leaves of the spinach.

ATTAGENUS MEGATOMA (*Fabr.*).

Examples of this beetle, one of the destructive family of *Dermestidæ*, have frequently been sent to me, as having been captured under and about carpets, with request for information of its character and habits.

Appearance. — It is an inconspicuous, small, black, shining beetle, about one-eighth of an inch long, twice as long as broad, quite regularly oval in outline, its small head bent downward and but partially seen from above. Its antennæ terminate in a large ovate club. The short legs and the body beneath are brown.

The larva, etc. — Although the beetle is a destructive household pest, yet it is without a common name. It has but lately come into notice, and its habits are not fully known. Associated as it is, in our houses, with the carpet-beetle, *Anthrenus scrophulariæ* (Fabr.), there is good reason to believe that the two are co-workers in mischief. Its larva is long and slender in comparison with that of the carpet-beetle, being about three-tenths of an inch in length. It is clothed with short brown hairs somewhat appressed to the surface, and terminates in a long pencil of hairs. This brief description is from memory as no example of it is at hand at the present writing.

Eats carpets. — The larva has been taken by me from beneath carpets and fed in confinement upon pieces of carpet, until it had completed its moltings and transformations and appeared as the perfect beetle. The insect appears to be on the increase, for not only have numerous examples occurred in my own residence during the month of

Jun:, but it is being frequently sent to me by correspondents from widely-separated localities.

The beetle a pollen feeder. — Like the carpet-beetle, the perfect insects find at least a portion of their food in the pollen of flowers, as I have captured them in large numbers in Washington park, Albany, on the flowers of species of *Spiræa*, associated with the *Anthrenus*. The beetle, after it has done all the harm that it can in providing for a continuation of the injuries of its larvæ, by depositing its eggs upon our carpets or other woolens, then displays a desire to gratify its harmless appetite upon pollen. Seeking to leave our rooms in search of food, it flies to the windows, and if these be closed, numbers of them may be taken from the lower portions of the window casings and of the window sashes. During the month of June, frequent examinations of the borders of the carpets along the sides of the rooms in which they are known to occur should be made for these little beetles, which, when found, may, with much satisfaction, be crushed by the finger nail. Those found upon the windows should also be killed, as they may possibly not have made deposit of their entire quota of eggs.

Infests hair-cloth furniture. — From examples of the beetle which have been brought to me by a lady in Albany, and from the statement made of the conditions under which they were found, it is quite probable that it breeds to quite an extent in hair-cloth furniture, as chairs, sofas, etc. One of its congenors, *Attagenus pellio* (Linn.), which received its specific name from its fondness for dried skins, and which occurs in the United States and over most of the civilized world, displays quite a varied taste for food, for it is stated of it that it also eats cotton and linen fabrics, and is sometimes quite injurious to carpets. A writer says of it: "I have known it to select a particular stripe, especially one of red flannel in the domestic fabric known as rag carpets, and follow it out into the middle of the room, gnawing it off at intervals."

Possibly eats cotton and linen fabrics. — The carpet-beetle, *A. scrophulariæ*, has been charged with injuring lace curtains trailing upon the carpets, but from all that we know of its habits, we are unwilling to accept such statements unless they shall be verified; we believe that it only feeds upon woolen material. It is not at all improbable that we have the author of these reported injuries to cotton and linens in the *Attagenus megatoma*. If this suspicion is hereafter confirmed, and its range of food found to embrace hair, furs, cotton, linen and wool, then it is unquestionably a pest more to be dreaded in our homes than the rapacious and destructive carpet-beetle.

Remedies. — Benzine or kerosene oil will kill the eggs, larva and pupa of this insect whenever they are brought in contact with them. It should, therefore, be freely used in their haunts underneath base-boards and the floor-joinings adjacent.

Preventives. — A material which promises to be a preventive of a repetition of the attack, if the insects be once destroyed, is the common roofing-paper prepared with gas-tar. It should be cut in strips of from eighteen inches to two feet in width, and placed underneath the border of the carpet. The odor that would penetrate heavy carpets should not be very disagreeable, and it certainly may be endured if it shall give us immunity from two or more serious carpet pests. Even thick paper or newspapers spread underneath carpets give a certain degree of protection, and none should be laid without such precaution.

MACRODACTYLUS SUBSPINOSUS (*Fabr.*).

A correspondent, Mrs. Lucy G. Chrisman, writing from Warren Farm, Chrisman, Rockingham Co., Va., gives as the result of her observations upon this insect, the well-known rose-beetle, that its depredations in that region are limited to localities having a sandy soil. Where clay occurs, the insect is not to be found. Should this statement, which, from its detail and accompanying facts, is an interesting one, be confirmed by observations throughout other sections where it abounds, it will prove to be an important portion of its life-history. It would enable us by devoting our clay soils, so far as it may be practicable, to the culture of the crops that are known to be the favorite food of this pest, to do much toward a mitigation of its ravages. The following is the statement:

Our county is a mixed sandy soil on the rivers, creeks and all the little streams, with clay land in many places between. Nowhere on the clay lands are the rose-bugs ever seen. I have a friend whose land lies on both sides of Cook's creek, and his sons' houses are quite near each other, but upon different sides of the stream; the one never sees a rose-bug; the other is eaten up with them. A mile distant is the town of Bridgewater, on a piece of North river bottom-land. In 1882, I was there in the rose-bug season, and the creatures filled the air so as to make riding exceedingly disagreeable — at times looking almost like a brown cloud — extending down to the water's edge, yet just across the river not one was to be seen. We seemed to have left them at the bridge in crossing.

A comparative exemption of a clayey district from the presence of this insect might naturally be expected from the difficulty that the female beetle would encounter in digging into the tough soil for the deposit of its eggs. But that this explanation does not meet the conditions as above given by our correspondent, appears from the statement

that she gives, that " the beetles came to us from toward the mountains
in swarms, and yet on one side of a little narrow creek there are none,
while on the other [sandy] side, they are abundant; and the same in
many places where no creek intervenes between the sand and the clay."
Their apparent dislike to clay land was so noticeable, that Mrs. Chrisman had thought of having some clay distributed over her rose-beds to
see if it would not act as a repellant.

INCREASE OF THE BEAN-WEEVIL.

Mr. Joseph Barker, of Marathon, Cortland county, N Y., states that
the bean-weevil, *Bruchus obsoletus* Say, is becoming quite troublesome in
his vicinity.   It infests field-beans (the small yellow bean), and it is
thought that persons have been sickened by eating of the beans.   Three
miles from Marathon, at Killoway Station on the Syracuse and Bing-
hamton Railroad, entire fields of beans are reported as having been
destroyed by the insect.

This insect was first noticed about twenty-five years ago in the New
England States.   It has since been of common occurrence in the State
of New York, has frequently shown itself in Pennsylvania, and occasionally in localities in the Western States as far west as in Missouri.

Although it is now known in various parts of the United States, it is
as yet confined to certain localities.   Every effort should therefore be
made to prevent its general distribution.   As some of the beetles do not
emerge from the beans until spring, they are liable to be planted with
the seed-beans, and the evil may thereby be continued and increased.
If the beans intended for seed be tightly tied up in stout paper bags and
be kept until the second year, there will then be no living beetles within
them, and they will be equally valuable for seed.   If, however, they have
been badly perforated they should not be used for planting, as many of
them would not germinate.

OVIPOSITION OF MONOHAMMUS CONFUSOR (*Kirby*).

Dr. Packard has remarked of this beetle — the long-horned pine borer,
that " it appears early in June and is to be found through the summer
until early in September; and at any time in July and August, as well
as the first week in September, it lays its eggs."

It was my fortune a number of years ago (in 1857) to have an excellent opportunity of observing the oviposition of this beetle.   My record
made at the time, differs in some particulars from that given by Dr.
Packard — perhaps owing to the different conditions under which oviposition was made:

7

September 9th.   Numbers of *Monohammus titillator* [*M. confusor*], were seen upon the trunks of a few pine trees [probably not exceeding ten inches in diameter] that had lately been felled upon the border of the pine grove at the cemetery [at Schoharie, N. Y.]. It was apparently, just their season for mating, for as many as ten or fifteen pairs could be counted upon a single trunk, at a time.   Rarely would a single female be found unattended, and the few such that were noticed, remained so but for a short time.   If during the coupling of a pair, a male chanced to discover them, a combat immediately ensued between the two males which would be fiercely contested and usually end in the interlocked combatants falling together to the ground.   Upon separating and the victor returning for his prize, he might have again to contest possession with another rival.   Some of the females, while still held by the male, were engaged in gnawing holes in the bark into which to place their eggs.   The beetle reversing her position, and thrusting the tip of her abdomen into the excavation, deposited her egg between the bark and the wood,* as shown by examples in my collection.   The egg is white, cylindrical, with rounded ends, of a length three times its breadth, and is invariably placed longitudinally with the tree.†

FIG. 2.— *a, a, a a*. Cuttings made in a fir tree by MONOHAMMUS CONFUSOR for oviposition : *b*, one laid open to show the egg and its position — natural size.

* Upon raising pieces of the bark into which I had seen the eggs placed, the eggs usually came away with them, lying exposed, with a large section (nearly one-half) of their longitudinal diameter showing above the (under) surface.   The eggs that Dr. Packard saw the beetle deposit in the fir were inserted *within the bark*, and the larvæ hatching from them, a week after oviposition "had begun to descend slightly *into* the bark" (*Ann. Rept. Dept. of Agriculture* for 1884, p. 381).

† In the accompanying figure by Dr. Packard, the cuttings for the eggs are represented as made indifferently either perpendicularly, obliquely or horizontally.   In this instance they were made in a living and standing fir tree; in my observations, in felled pine trees.

An egg having been placed in position, the pair, if not meanwhile disturbed, would again unite *in copula*, after which another cavity would be made and another egg deposited, as before. Query: are repeated coitions necessary to the fertilization of the eggs, and do they require to be separately fertilized?

Another note made by me July 23, 1868, records the same beetle in abundance at Schoharie, N. Y.; on the branches of some young pines — the sexes in coitu.

OTIORHYNCHUS LIGNEUS INFESTING A DWELLING-HOUSE.

This curculionid beetle was received from Dr. C. M. Coe, of Lycoming, Oswego county, N. Y., in June, 1884, for identification, with the statement that upon opening his house in the month of May, after it had been unoccupied for four years, it was found to be overrun with this insect.

Upon communicating to Dr. Coe the name of the beetle, and expressing my inability to account for its infesting his house in such numbers as stated, since nearly all of its allied species are vegetable feeders, and asking for additional information of its occurrence, he kindly sent me the following items:

The beetles were first noticed upon opening the house in the month of April, when they were found in large numbers in every part of the house, both dead and living. The living ones continued to abound without apparent diminution in number until about the first of June, since which time they have been steadily disappearing. No measures were taken for destroying them beyond a daily sweeping of the rooms and burning such as were collected. At first, as many as two quarts, it was thought, could be gathered at a sweeping. At the date of the letter written (July 10th) not one was seen where previously a hundred had been observed. Nothing had been left in the house at the time of its closing, upon which, so far as known, the beetles could have bred or fed, if attracted thither from out of doors.

As the larvæ of these species of *Otiorhynchus* so far as known live upon plants, and for the most part upon their roots, we have no explanation to offer for the remarkable abundance of this species in the instance above mentioned. *Otiorhynchus sulcatus* (Fabr.), a species introduced from Europe, has long been known to be injurious to many of the garden products in both its larval and perfect stages. As a larva, it is often quite destructive, in Europe, to the leaves of strawberries, but has not yet been known to attack them in the United States. Mr. S. Henshaw reports this species as injurious to bulbs and house-plants. In Europe it injures leaves and roots of grapevines (*Ent. Month. Mag.*, xii, 83).

Following the reception of the *O. ligneus* from Dr. Coe, Prof. Cook of Lansing, Mich., informed me that the same species, long known as a common insect, had just been discovered as injurious to the strawberry.

O. picipes has of late years been quite destructive to raspberries, in Europe, destroying in one instance, two acres of the plants, in Cornwall. The weevil strips the leaves from the plants, destroys the tender shoots, and eats the bark from the canes (*Amer. Entomol.*, iii, 1880, p. 127). Dr. Packard has recorded (erroneously?) its occurrence in Massachusetts.

For attack of *O. sulcatus* and *O. picipes* upon the raspberry and Primula see Miss Ormerod's Report for 1879, pp. 6, 7; Report of 1880, p. 4, on vines.

O. tenebricosus, of Europe, feeds on the buds, young shoots, bark and leaves of the Apricot, Nectarine, Peach, and Plum. In the larval state it injures the roots of raspberries, gooseberries, strawberries and garden vegetables (*Miss Ormerod's Manual*, p. 306).

SPHENOPHORUS SCULPTILIS *Uhler*.

In the notice of this insect in the *First Ann. Report N. Y. State Entomol.*, 1882, pp. 253–263, it is stated (p. 257) that it seems not to have been observed in the State of New York between the years 1861 and 1867. The following item, from the *Rural American* of July 15, 1866, appears in the *Practical Entomologist*, for November, 1866 (vol. ii, p. 20), from a correspondent in Hannibal, N. Y.:

MR. MINER: — Knowing that you are interested in any thing connected with agricultural pursuits, I take the liberty of sending, for your inspection, a few specimens of small beetles, taken out of three hills of corn. They burrow down in the hill, and attach themselves (head downward) to the young corn, about two inches below the surface of the ground, and insert their proboscis into the corn plant, and suck the juice until the blade turns blue and dies. I find from one to five of them in each hill. One of my neighbors has lost eight acres of corn (old sheep pasture) by them. If you can suggest any thing to stop their ravages, you will confer a favor on several subscribers to the *Rural American.*

The insects were not identified, the editor merely remarking that the "small beetles sent to us are an insect with which we are not acquainted." Mr. Walsh in copying the article, regrets that the correspondent gave no other characteristic of the beetle except that it was "small," and that the editor had not referred to him or some other entomologist for its name. It is very probable, from the habits given and the locality that the beetle was *Sphenophorus sculptilis*, which the following year was described as *Sphenophorus zeæ*, by Mr. Walsh, from examples received from Onondaga county. Hannibal is in Oswego county, which borders Onondaga county on the north.

In treating of the larval life of this insect in the Report above cited, I wrote as follows:

I can see no reason for supposing that the larva of this species should so far depart from what is known of the habits of the family, as to feed upon decaying wood. Westwood says of the *Curculionidæ*, " these insects are entirely herbivorous, some feeding upon leaves, others upon seeds, and some upon the stems of vegetables." Riley asserts (*Third Missouri Report*, p. 10), of the members of the same family, " with the exception of an European species (*Anthribus varius*) whose larvæ were found by Ratzeburg to destroy bark-lice, they are all vegetarians, the larvæ inhabiting either the roots, stems, leaves or fruits of plants, and the beetles feeding on the same."

Referring to the above paragraph, Mr. Warren Knaus, of Salina, Kansas, has sent me the following communication:

" I see that you do not indorse the idea that any of the curculio larvæ are lignivorous. I have never observed the habits of *Sphenophorus sculptilis*, having only taken two or three specimens in this county, but for the past three seasons I have taken *Wollastonia quercicola* (Boheman) from decaying cottonwood [*Populus monilifera*] logs and stumps, and have never taken them in any other locality. I have observed logs perforated in every direction, and have found small white larvæ at work in these logs, and from the same logs have taken the above-named perfect beetle. I also find the beetle in the larval burrows, but have not observed their habits sufficiently to determine definitely whether these larvæ are those of *W. quercicola*. I trust, however, that next season I may be able to test the truth or falsity of my present belief founded on observations, that the larvæ of this genus are lignivorous."

DESTRUCTION OF SPRUCES AND FIRS BY BARK-BORERS.

Extensive destruction of the spruces (*Abies nigra* and *A. alba*) and firs (*Abies balsamea*) through the ravages of bark-boring beetles has for several years past been observed in Northern New York and New England.

The attention of Prof. C. H. Peck, N. Y. State Botanist, had been called, in 1873, to the fact that in some parts of the Great Northern Wilderness of New York the spruce trees were rapidly dying, to the great pecuniary loss of the lumbermen and land-owners. In some tracts of considerable extent, nearly all the spruces were reported as having been killed, giving to the forest a prevailing brown hue as if a fire had run through them. None of these affected districts, however, came under the observation of the State Botanist at that time (27th *Ann. Rept. on the N. Y. State Museum of Natural History*, 1875, p. 75).

The following year, Prof. Peck reports, that the dying of the spruces was not of recent origin, but that it had been known in Lewis county from ten to fifteen years previous, and in Rensselaer county, the same destruction had been observed as early as the year 1845.

A locality near Lake Pleasant, Hamilton county, where the spruces were rapidly dying, was visited by him. As the result of a careful investigation made it was found that their death was owing to the countless winding galleries made by one of the bark-mining beetles, *Hylurgus* [*Dendroctonus*] *rufipennis* Kirby, between the bark and the wood. A part of each was eaten by the insect, consisting of the newly-formed and forming layers of wood and bark the most vital parts of the tree. Its operations were therefore equivalent to a girdling of the tree. In one instance, another of the bark-borers, of a much smaller size, *Apate* [*Polygraphus*] *rufipennis* Kirby, was found associated with *H. rufipennis,* in its destructive operations.

The report embraces an interesting discussion of the above attack, in its character, localities of its occurrence, frequency with which the larger trees show the attack, its cessation in some localities without apparent cause, remedies for it, etc., for which the 28th *Rept. of the N. Y. State Museum of Natural History*, 1879, pp. 32–38 may be consulted.

In the 30th *Report of the St. Museum of Natural History*, 1878, pp. 23–25, Professor Peck has given additional observations on the spruce attack by *Hylurgus rufipennis* and presented details of the operations of the beetle in the construction of its burrows.

In his following Report, Prof. Peck describes an attack upon the balsam fir, *Abies balsamea*, by another bark-boring beetle, observed by him at Summit, Schoharie county, N. Y., through which a number of trees had been killed and others were dying. The burrows were carried underneath the bark in a horizontal direction, so that three or four occurring at about the same height in the trunk would completely girdle the tree and destroy its life. The beetle proved to be a *Tomicus* of probably an undescribed species (31st *Rept. of the N. Y. State Museum of Natural History*, 1879, pp 22, 23).

The following determination, contained in a letter received by me from Dr. LeConte, in December of 1877, is believed to refer to the above insect: "*Tomicus,* related somewhat to the European species, *suturalis, curvidens* and *laricis* — perhaps imported." Dr. LeConte desired additional specimens "of the male with the hairy head " to be sent to him, which probably for some reason, was not done.

Dr. A. S. Packard, in his *Insects Injurious to Forest and Shade Trees,* 1881, pp. 219–227, has written of the destruction of spruces and firs in Northern New England in 1878–1881, and states, as the result of his observations and of reports made to him, that the destruction of the spruces was chiefly owing to three species of the cylindrical bark-boring *Scolytidæ*, viz., *Pityophthorus puberulus, Xyloterus bivittatus,* and *Xyleborus cælatus.* These, aided by *Monohammus confusor,* were also found to have caused the death of large and healthy firs, a foot in diameter.

In the *Report of the Commissioner of Agriculture* for the year 1884, Dr. Packard has presented a "Second Report on the Causes of the Destruction of the Evergreen and other Forest-trees in Northern New England and New York." In the Adirondacks of New York, he had found many dead spruces and firs with their "bark filled with species of *Dendroctonus* and *Tomicus*, or allied genera. Living spruces with the leaves fresh and green, contained in their bark, in June, the larva and the beetle of *Hylurgops*, running their burrows so as to girdle the trees. In Northern Maine the destruction of the spruces was still continuing, but was apparently abating.

Remedies. — As remedies for the ravages of these bark-borers, Professor Packard has proposed: Stripping the dried bark from infested trees and burning it. Cutting down dead trees and disposing of them for fuel. Stumps remaining from trees recently cut, should be barked, and the bark burned.

In the Harz forests in Germany, where one of these bark-borers, *Bostrichus typographus*, has occasioned such enormous losses, causing in the years 1780–1790, the death of two million of trees, other methods for arresting its continued attack have lately been resorted to, under the direction of the Forestry Commission. They are thus stated:

Experienced men are told off, to go through the forest and search for the trees attacked by the beetle, and fell and bark them to prevent the spreading of the insects. In most cases they are quite able to hold the insects in check. These generally attack trees loosened in the roots by wind, known after the beetle gets in by their foliage turning yellow. In spring, when they are worst, *healthy living trees are felled at the southern margin of the forest in many spots, for the purpose of attracting the beetle.* Such trees are often full of them three or four days after being felled. The trees attacked are barked, which destroys the larvæ if not too far advanced; if so, the bark is burned. To prevent any escaping while barking, a cloth is spread under the stem.

The above method, it is believed, could be employed with great benefit for the prevention of our spruces and firs in infested districts. The attraction that newly-felled trees have for many of the bark and timber-boring beetles at the season of their oviposition has often been recorded in our entomological literature. Note has been made in a preceding page of the numbers of *Monohammus confusor* that were drwan to some pine-trees for oviposition.

DISTRIBUTION OF THE HARLEQUIN CABBAGE-BUG.

In the 1st Annual Report of the State Entomologist some speculations upon the probable future extension of the range of *Murgantia histrionica* into New York, the Eastern States and the Dominion of

Canada, were indulged in, based upon its observation in Denver, north of the isothermal line of 40° Fahr.

The following observations of it, by Mr. A. S. Fuller, of Ridgewood, N. J., kindly communicated to me, would indicate a power of endurance of cold possessed by it, to a far greater degree than hitherto supposed:

" I have seen it by the millions feeding on sunflowers in New Mexico, at an elevation of 9,000 feet, where a temperature of 30° below zero in winter is not at all unusual."

DACTYLOPIUS LONGIFILIS *Comstock.*

This Coccid was received July 8, 1882, from Mr. George B. Simpson. It occurred in Waterbury, Conn., on the prickly-pear cactus, where it was first observed the previous winter. Mr. S. states that the eggs seem to be placed, in preference, on the strips of cloth with which the plants are tied to the stakes.

For the description and illustrative figures of this species, see Prof. Comstock's Report in the *Report of the Commissioner of Agriculture* for the year 1880, p. 344, pl. 11, f. 2, and pl. 22, f. 1.

HELIOTHRIPS HÆMORRHOIDALIS *Bouché.*

In my First Annual Report on the Insects of New York, in the list of Insect Depredators upon the Apple-tree, the above species was included upon the statement of its observation by Mr. Theodore Pergande, of the Entomological Division of the Department of Agriculture, as follows:

"This year [1882], as late as November 14, after several quite cold days, I found for the first time *Heliothrips hæmorrhoidalis* Bouché, on apple leaves in the orchard of the U. S. Department of Agriculture, as lively and active as in hot-houses, where this species was only observed previously."

The special study which Mr. Pergande is giving to the *Thripidæ*, an interesting and but little known family of the Homoptera, renders valuable the following note from him in relation to the above insect:

"I wish to call your attention to the fact that *Heliothrips hæmorrhoidalis* is not strictly an insect which depredates on the apple, but which was only accidentally on the trees, having evidently wandered from the hothouse plants near by to the apple, and was not again seen on the trees the last year. If the climate would be constantly warm, I have no doubt but that it would live and multiply on the apple as well as it does on a great variety of plants which are kept in hot-houses. I simply mentioned it to show how readily foreign species of insects may be introduced, if they find the climate and other conditions congenial to their propagation."

INJURIOUS LEPIDOPTEROUS INSECTS.

Melittia cucurbitæ Harris.

The Squash-vine Borer.

(Ord. LEPIDOPTERA: Fam. ÆGERIADÆ.)

Ægeria cucurbitæ HARRIS : in New England Farmer for Aug. 22, 1828, vii, No. 5, p. 33; Amer. Journ. Sci.-Arts., xxxvi, 1834, p. 310–11 ; Treat. Ins. N. Eng., 1852. p. 252–3 ; Ins. Inj. Veg., 1862, p. 331, figs. 159, 160, pl. 5, f. 8 ; Entomolog. Corr., 1869, p. 284–5.

Ægeria cucurbitæ. RILEY : 2d Rept. Ins. Mo., 1870, p. 64.— REED : in Rept. Ent. Soc. Ont. for 1871, p. 89–90, figs. 96–98.— THOMAS : 6th Rept. Ins. Ill. [1877], p. 41.— FRENCH : in 7th Rept. Ins. Ill., 1878, p. 173.— MARTEN : in 10th Rept. Ins. Ill., 1881, p. 107 (larva).— SAUNDERS : Ins. Inj. Fruits, 1883, p. 361–2, figs. 370–372.

Melittia cucurbitæ PACKARD : Guide Stud. Ins., 1862, p. 279, f. 210 ; in Hayden's 9th Rept. U. S. Geolog.-Geograph. Surv. Terr., 1877, p. 769, f. 36.— COOK: in 13th Rept. St. Bd. Agricul. Mich., 1875, p. 116 (in Mich.).— COLEMAN: in Papilio, ii, 1882, p. 50 (larval hibernation).— HULST: in Bull. Brookl. Ent. Soc., vi, 1883, p. 10 (habits, etc.).—LINTNER : in Count. Gent., xlix 1884, pp. 477, 487, 517.

Melittia Ceto GROTE : New Check-List N. A. Moths, 1882, p. 10.

A correspondent of the *Country Gentleman* has sent to that journal the following communication, telling of the ineffectual efforts made by him and his neighbors in dealing with the squash-vine borer, and asking for information of the insect, and approved means of checking its destructive attacks. Similar appeals and inquiries from several other localities have shown an urgent need of the information asked for, in view of a widespread extension of its injuries, while at tne same time assuming a more serious character than in the earlier years of our acquaintance with the insect :

EDS. COUNTRY GENTLEMAN — I have read in various numbers of your paper statements of crops of Hubbard squashes and prices obtained for them. In behalf of a large number of your subscribers in the valley

8

of the Hudson, I ask the State Entomologist to give us the full history of an insect that stings the squash vines after they have commenced to run, and lays an egg in the puncture, which shortly develops into a maggot or grub, 1 1-4 to 1 1-2 inches long, that soon, if left alone, eats and bores the vine so as to destroy all its substance, and of course kills it. We would like its full history; its scientific name; how and when it breeds ; what changes it undergoes after the boring stage, in which they make such havoc in squash vines; what the perfect insect looks like, and above all, how we are to best meet it. Every one here has given up raising the Hubbard squash, and the insect has gone to laying its eggs in the pumpkin vines for want of something better. Every remedy tried has failed — even going over each vine daily and cutting out the parts where the insect lays its eggs. For the last crop I tried to raise, I spent nearly $100 in labor, going over the vines each day and cutting out the grubs. I should not have grudged this if it had saved the crop, but it did not. I lost every vine, and did not have a single squash.

We have never tried poison, as the borer is in the very center of the vine, and I do not see how the poison can reach it. It is possible that the State Entomologist can suggest something that would put an end to the pest. To give some idea how bad it is, I would say that I have cut out one hundred and forty-two grubs from a single vine, though not at one time. , I have lately thought of making a box six feet square, and covering it tightly with a mosquito netting, and using this to cover each hill. It will be costly, but if effective I could stand it, for I do not like to have an insect get the best of me, and deprive me of so delicious a vegetable. I have seen complaints of these borers in your paper, and have seen it ascribed to a dozen different insects. Therefore we look to the State Entomologist to give us the truth.   P. V. B.
Coxsackie, N. Y.

The insect of which inquiry is above made, is the one commonly known as the " squash-vine borer." It was described by Dr. Harris, from the moth, as _Egeria cucurbitæ_. Later it has been transferred to the genus _Melittia_, and as _M. cucurbitæ_ it has been frequently referred to by recent writers. In a check-list of North American moths, by A. R. Grote, published in 1882, it is regarded as identical with a species described by Westwood of England prior to Dr. Harris' description, under the name of _Ceto_, but until its identity is more clearly shown, it may better retain its well-known and appropriate name of _cucurbitæ_. Fig. 1 represents the moth (a male) one-half larger than the natural size.

Description of the Moth.

The earliest description of the species is that of Dr. Harris, in the _New England Farmer_ for August 22, 1828. It is of the moth only, and is as follows :

" Body tawny, with four or five black dorsal spots; anterior wings olivaceous brown; posterior wings, except the margin and nervures, hyaline; tibiæ and tarsi of the hind legs densely fringed with fulvous and black hairs. Length of the body half an inch. The wings expand one inch and one-quarter." Later (*Silliman's Am. Jour. Sci.-Arts*, xxxvi, 1834, pp. 310–311) Dr. Harris gives the following somewhat fuller description : "The wings opaque, lustrous olive-brown ; hind wings transparent, with the margin and fringe brown ; antennæ greenish black; palpi pale yellow, with a little black tuft near the top; thorax olive; abdomen deep orange, with a transverse basal black band, and a longitudinal row of five or six black spots; tibiæ and tarsi [shanks and feet] of the hind legs thickly fringed on the inside with black and on the outside with long orange-colored hairs; spurs covered with white hairs. Expands from thirteen to fifteen lines."

Fig. 3.—Moth (a male) of the Squash-vine borer, MELITTIA CUCURBITÆ, enlarged one-half from the natural size.

In Dr. Harris' latest publication (Treatise on Insects Injurious to Vegetation), only the more salient features of the moth are given. It is referred to, as "conspicuous for its orange-colored body spotted with black, and its hind legs fringed with long, orange-colored and black hairs. The hind wings only are transparent, and the fore wings expand from one inch to one inch and a half." The colors as given above are subject to considerable variation. Mr. Hulst states : "The ordinary orange color is more marked in the female than in the male. One female had the body almost wholly black. In some specimens yellow takes the place of orange, and in one fresh male the abdomen was almost white, and the fringes of the legs, ordinarily orange, were very light yellow."

Description of the Larva.

In none of the editions of the Harris report is the caterpillar which develops into the moth described — a strange omission, since it is in this stage that it is met with by the squash-grower hundreds of times more frequently than is the perfect insect — the popular remedy against the insect being the cutting out of the caterpillar from the vine. In Dr. Harris' Entomological Correspondence, published in 1869 (pages 284–5), it is, however, described minutely, as follows :

"Aug. 15, 1841 — Fully grown larva. Somewhat depressed, fleshy, soft, tapering at each extremity ; segments ten in number, very distinct, the incisions being deep; the eleventh or last segment minute, and hardly distinct from the tenth. Head retractile, small, brown, paler on the front, and with the usual V-like mark on it. First segment or collar, with two oblique brown marks on the top, converging behind. A dark line, occasioned by the dorsal vessel seen through the transparent skin along the top of the back, from the fourth to the tenth rings

Fig. 4.—The Squash-vine borer, MELITTIA CUCURBITÆ.

inclusive. True legs six, articulated, brown ; prolegs wanting, or replaced by double rows of hooks in pairs beneath the sixth, seventh,
eighth and ninth rings, and two single rows under the last rings. Spiracles brown. A few very short hairs on each ring, arising singly from
little hard points or pit-like, warty substances. Length from one inch
to one inch and a fourth."

This is followed with the statement that on "August 17th, one of
these formed its cocoon of fragments of squash-stalks tied together with
a few silken threads." It is preceded with the mention: "Found summer and winter near the roots of squash vines, and also
in the roots "— probably referring to the cocoons. The
larva is shown in Fig. 4, and in Fig. 5, the cocoon containing its pupa. Those who have access to the edition of Harris' "Insects Injurious to Vegetation," containing colored plates (published in 1862), may find there (pl. V, fig. 8)
a figure which will enable them readily to recognize the insect when it
is seen, during the month of July, flying over the plants, and alighting,
from time to time, to deposit an egg. To others it may be of service
to state, in addition to the above description of the moth, that the insect
belongs to the family of *Egeriadæ*, sometimes known as "clear-wings,"
from the wings being without scales. In this species, however, only the
hind wings are transparent, the front ones being opaque, and covered
with scales, as in nearly all of the *Lepidoptera*.

FIG. 5.—Cocoon show-
ing the pupa of the
Squash-vine borer.

The Family of Ægeriadæ.

The *Egeriadæ* bear a marked resemblance to wasps and hornets,
with their orange markings, narrow and usually transparent wings, and
their flight by day. Most of them have
at the end of the body a tuft or brush of
stiff hairs, which they can spread out, at
pleasure, like a fan. The currant-stem
borer (*Egeria tipuliformis*) illustrated in
Fig. 6, and the peach-tree borer (*E. exitiosa*) are perhaps the best known species
of the group. Within a few years past
the number of species of this interesting
family of internal borers in plants, shrubs
and trees, has been largely augmented,
through the studies of Mr. Henry Edwards, of New York. The latest
check-list gives one hundred and eleven North American species, arranged in seventeen genera, while Dr. Harris' Catalogue of North
American Sphinges, published in 1834, named but twelve species.

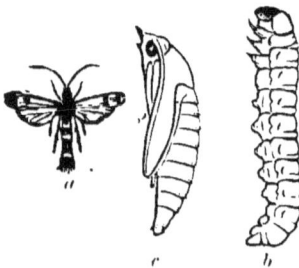

FIG. 6.—The Currant-stem borer, ÆGERIA
TIPULIFORMIS; *a*, the moth, the natural
size; *b*, the larva, and *c*, the pupa, both
greatly enlarged.

Life-History and Habits.

Our knowledge of the several stages of the life of this pest is quite incomplete, as we find when we attempt to indicate the best methods for preventing its injuries.

In the New England States the parent moth, according to Dr. Harris, may be seen flying about the plants from the 10th of July until the middle of August, attaching its eggs to the vines close to the roots. Its habits have been carefully observed by the Rev. Mr. Hulst of Brooklyn, N. Y., to whom we are indebted for the following statement: "The moth appears on Long Island shortly after July 1st [and probably earlier.] During the summer of 1882, I captured some thirty specimens about a small bed of summer squashes in a neighbor's garden. The moths fly during the day, being the most active during the hottest sunshine, and quiet in the early morning. I have seen only two pairs mated, and this was between 2 and 3 P. M. The female lays her eggs morning and afternoon, mostly on the stalk of the plant just below the ground. She extends her abdomen into the crack of the ground about the stem of the plant, and the most of the eggs that I have seen were from one-fourth to half an inch below the surface. Often, however, they were laid a foot above the ground, and in a few instances were observed upon the petioles of the leaves."

The egg is oval and of a dull red color. The length of time required for its hatching is not known, but it is probably about a week. Upon hatching, the young larva at once burrows into the stem. It grows rapidly, and when about half-grown, its effects are visible in the wilting of the vines. Later, as the attack continues, and several larvæ unite in it, the vines die down to the root. This, in the latitude of New York, is usually in the month of August.

About the first of September some of the larvæ have attained their full growth, when they escape from the vine and construct their cocoons a little below the surface of the ground in which to undergo their subsequent changes, first to the pupa and lastly to the imago or moth. It is possible that the larva does not always enter the ground for its final transformations. Dr. Harris distinctly states that the cocoon is formed in the ground of earth cemented by a gummy matter, but Dr. Packard records of the larva that "it lives in the vine (in New England) until the last of September, or early in October, when it either deserts the vine and spins a rude earthen cocoon near the roots, or, as is often the case, remains in the hollow it has made in the stalk, and then changes to a chrysalis."

Mr. Henry Stewart, of Hackensack, N. J., who has made a study of the insect, says that he has never found the cocoons within the vines and that they do not pupate therein.

In the Southern States the more advanced larvæ mature and form their cocoons as early as in the month of August. Of some examples received by me from Baltimore, Md., one of them spun itself up in a silken cocoon in an angle of the box in which it had been placed on the day of its reception, August 14.

Until recently it has been supposed that the larva changed to a pupa shortly after it had made its cocoon, in accordance with what has been observed in the transformations of by far the larger portion of our insects. More careful observations seem to show that the larva, usually at least, continues in its caterpiller state throughout the winter. Mr. N. Coleman, of Berlin, Conn., under date of February 24, has given the following statement:

" From numerous observations, I am assured that this insect hibernates in the caterpillar stage, and does not transform to the pupa state until the spring. The pupa cases [cocoons] are formed in the latter part of the summer, and, in every instance thus far, I have found the larvæ in the cases unchanged. The last examination was made only a day or two since."

With the emergence of the moth from its pupal case early in July, as previously stated, its appearance on the wing, the coupling of the sexes and the deposit of the eggs, its life cycle is completed.

Its Injuries.

The reports of injuries by this insect are becoming more frequent, and are assuming a more serious form than heretofore. It was formerly supposed that the earlier squashes were the more liable to be infested by it, but of late the Hubbard suffers the most severely, and to an extent that in some localities is preventive of its culture. The statement given on a preceding page by P. V. B., of Coxsackie, of the effort made to repel the attack, the expenditure involved, the cutting out of *one hundred and forty-two larvæ from a single vine*, and the entire subsequent loss of the crop, tells more fully than has been told before, the story of the harm that this pest is now inflicting, not only along the Hudson river, but in some of the Eastern States, New Jersey and elsewhere — fortunately not to as great an extent, as yet, in our Western States. The gentleman, in a subsequent communication, informs me that it has been only known in his vicinity for the last twelve or fifteen years, and during that time he has taken thousands of the borers from the vines, in sizes varying from newly-hatched to one inch and a half long.

In New Jersey the insect seems to abound, and extensive injuries are reported. A gentleman from Hackensack writes: "I have taken over a hundred of the borers from one squash vine in the past few days [in August], finding them in different parts, from the roots to the end of the vines."

Although reported occasionally as a borer in pumpkin vines, it is not known as the cause of serious harm to this plant. It also occurs in melon vines.

Remedies and Preventives.

Until the life-history of this species has been worked out with some degree of completeness — which can only follow many careful observations upon the habits and the several stages and conditions of the insect — can we be prepared to announce the best method of preventing its ravages. For the present, therefore, we must be content with giving such measures as promise material mitigation. Among those to be referred to, one or more, if carefully employed, will in all probability be found effectual in bringing the pest under control, so that entire crops of squashes will not be hereafter completely cut off; and complaints like the following (coming from Ulster county, N. Y.), need not be repeated: "I can fight off the striped bugs, but these little subterranean rascals are too much for me. I have frequently changed the location of my squash plot, and though occasionally managing to raise a small crop have oftener failed " (*Country Gentleman*, March 5, 1874, page 151).

First in order, those methods may properly be considered which may be employed at any time after the removal of a badly infested crop, with a view of preventing the recurrence of the attack the following year.

Crushing the larvæ by plowing, etc. — In the autumn, the entire field in which the borers have abounded should be plowed and harrowed two or three times. Such thorough working of the ground would crush the cocoons and their inclosed larvæ thrown up by the plow. It is believed that very few could by any possibility escape injury sufficient to prevent their final development. The depth of the plowing need not exceed the depth to which the larvæ bury themselves beneath the surface for pupation (probably not more than six inches ordinarily, but to be definitely ascertained hereafter).

Gas-lime, kerosene, etc. — A liberal distribution of fresh gas-lime, when procurable, after the removal of the crop, would kill the larvæ within the cocoons. Sprinkling the ground with kerosene oil before a heavy rain to carry it into the soil, might be as efficient as gas-lime. If mixed with soapsuds it might perhaps do quite as well. An infested crop should not be followed by another upon the same ground. A few hills might, however, be planted to serve as a decoy to concentrate the attack — the vines with their contained larvæ to be destroyed before the time for pupation. The parent moth seldom flies to any distance for oviposition, if the larval food plant is at hand.

Protecting by netting. — Covering with netting, as a method of preventing the deposit of eggs, by stretching netting (mosquito netting) upon a large frame, of perhaps eight feet square, over the plants, as has been suggested, I regard as well worth the trial. If the frames be used upon ground where the squash has not been previously grown, and where the insect is not already in the soil, it should prove an effectual preventive, if the eggs are all deposited near the root. But if oviposition is made, as has been stated by some writers, along the entire vine to its end, then the frame could prove but partially protective. Might it not be possible, by late planting, to limit the growth to the area of the frame until the time for the deposit of eggs shall have passed ?

Cutting out the larvæ. — Until recently, no other method of protection from this insect than that of cutting out had been suggested. It has been largely employed and, where the attack has not been severe, has proved effectual. A little practice renders one very expert in detecting the locality of the hidden depredator, in slitting the vine with the point of a pocket-knife and removing the larva. When, however, thousands have to be destroyed in order to insure the crop, the method is quite unsatisfactory. It certainly is not reliable under the great increase of the insect in recent years, as stimulated by the increased cultivation of the Hubbard squash.

Bisulphide of carbon in the ground. — In a paper read before the American Association for the Advancement of Science, at Boston, in 1881, Prof. A. J. Cook, of the Michigan Agricultural College, announced that he had found bisulphide of carbon to be an effective remedy against the squash-borer. The method was given as follows : "A small hole is made in the earth near the main root of the plant by the use of a walking stick or other rod, and about half a teaspoonful of the liquid poured in, when the hole is quickly filled with earth and pressed down by the foot. In every instance the insects were killed without injury to the plant." I regret to be obliged to state that the above claim has not been sustained by farther experiments. Prof. Cook, upon inquiry made by me, informs me that later trials with the material have failed to kill the borers, and he can no longer recommend its use for the purpose.

Rooting the plants at their joints. — Mr. Henry Stewart, of Bergen county, N. J., the author of many valuable notes on insect habits and depredations, has given the following as his method of dealing with the borer :

" Squashes in the East, and more particularly near the large cities where they are often largely cultivated for market, are infested with the squash-vine borer, the larva of an orange-colored moth (*Melittia cucurbitæ*), which not only deposits its eggs on the vines near the roots, as stated by Harris, but also, as I have often observed, along the vine here and there at the joints, the young larvæ entering the vine and burrowing it from end to end.   This destroys the vine and, of course, the crop, unless counteracted by some method.   The method of culture must be controlled by this danger ; and to manure the ground broadcast, as well as in the hill, and cover every joint when the vine begins to run, with soil raised by a hoe, avoids the danger, because the vines root at every joint when thus treated, and the original root may be cut off and yet the vine flourish all the more from its several rooted joints.   This cannot be well done with hill-manuring ; but yet I have found a way to circumvent the borer even then.   It is to procure Peruvian guano, or a special fertilizer prepared for squashes, and hill up around the joints as soon as the presence of the borer is perceived, or before that as a precaution, scattering a small quantity of the fertilizer (one ounce or so) about the newly earthed-up joint.   At the same time, a dash or two of water in which London purple has been mixed may be thrown on the vine near the covered joint with a small bunch of broom corn.   This will destroy many of the young larvæ as they eat their way into the vine, and the scent of the fertilizer may drive off the parent moth." (*Country Gentleman* for June 3, 1880, xlv, page 358.)

Treatment with saltpetre. — The following treatment of the vines with saltpetre has been recommended :

" Four tablespoonfuls dissolved in a pail of water, and about a quart applied to each hill where an attack was noticed and the leaves were wilting, at the time when the vines were just beginning to run nicely, effectually arrested the attack and a fine crop followed." (*Country Gentleman*, May 1, 1879, page 279.)

Use of Counterodorants.

In accordance with the views advanced by me and treated of at length in my First Annual Report on the Insects of New York, I believe that the deposit of the eggs upon the plant by the moth may be prevented by the use of strong-smelling substances, such as kerosene oil, coal-tar, tar-water, naphthaline, soluble phenyle, carbolic acid, and bisulphide of carbon.   Experiments are needed to determine the relative efficiency of the above and like materials.   It would, of course, be necessary that the applications should be renewed from time to time as their strength becomes impaired, so long as the moth continues abroad for laying her eggs — unfortunately a long period.

Efficacy of bisulphide of carbon. — I am quite desirous that faithful experiment be made with bisulphide of carbon, in consideration of its having proved effectual in preventing the attack of rose-bugs upon

9

grapevines, as narrated in the *Country Gentleman* of March 6, 1884, page 191. In that instance small vials containing a few drops each of the liquid were tied among the vines at intervals of a foot or two, the liquid being renewed every three or four days. As the result, grapes were grown upon the trellis as far as the vials were hung, while beyond, all the blossoms were eaten by the beetles. The cost of the bisulphide of carbon distributed in the twenty vials employed was less than twenty-five cents. The best method of using this material as a protection of squash vines, may safely be left to the judgment of the squash-grower.

Coal-tar, naphthaline, etc.—Coal-tar, as a preventive, might be dropped in patches upon the ground near the root and vines. Tar-water might be sprinkled over the ground. Pieces of cloth might be saturated with kerosene oil and wound upon sticks thrust in the ground, or sand moistened with it scattered upon the surface. Lumps of naphthaline would yield their strong odor for some time if lightly covered with the soil. Carbolic washes, if not too strong, could be sprinkled over the plants. Different methods of use of the above, as well as other materials, will undoubtedly come to the mind of one familiar with the cultivation of the squash.

Additional Information of the Insect Asked For.

That the remedies and preventives to be employed against this insect may be the best directed and the most effective, I would solicit the aid of all interested in the cultivation of the squash whom these lines may reach, in sending me information upon the following points. Each one should regard it as his duty to contribute, from his observations, to the extent of his ability.

Earliest date of the appearance of the moth. Latest that the moth is seen.

Is the egg ever deposited upon the root-stalk ? If so, how often ?

Greatest distance apart upon the vines that the egg is placed or the borer found.

Are the eggs placed on the upper or lower sides of the vines, or intermediately ?

Is the borer ever found in the main stalk ?

Earliest and latest notice of injury from the borer.

Earliest maturity of the borer and escape from the vine for pupation.

Are the cocoons ever found within the vines?

At what depth in the ground are the cocoons buried ?

Are the cocoons ever found far from the vines ?

Does the caterpillar ever change within the cocoon to the pupa during the autumn or winter? (In this stage it shows the legs, wing-covers and other organs of the future moth.)

What means have been found the most successful against the borer ? The number of points above given upon which information is desired, will serve to show that very little is known of this destructive insect and its habits, and that it is now proposed to give it the critical study that it has hitherto failed to receive.

Experiments at the State Farm.

Following the above request for additional observations and experiments upon this insect, Mr. E. S. Goff, horticulturist of the New York Agricultural Experiment Station, has very kindly communicated to me the gratifying results of his experiments upon it at the Station, both with insecticides and counterodorants, together with observations upon its habits, transformations, etc.

Under date of September 29, 1884, the following communication was received from him :

" It gives me pleasure to write to you that my experiments upon the squash-vine borer (*Melittia cucurbitæ*) offer much hope that we shall be able to battle this enemy with success. It has not been troublesome with us this season, though it has appeared in sufficient numbers to give an opportunity to show the value of my applications. I have just made a very thorough examination of the vines experimented upon, pulling up each plant and carefully looking over the stem. I found in seven hills of Hubbard squash, to which I had made no application, twenty borers (or cavities where they had been). In seven hills of the same, where I used a solution of copperas about the roots, I found eighteen borer cavities. In eight hills of the same variety I wet the stems a distance of two feet from the base of the plants with water containing Paris green at the rate of half a teaspoonful to a gallon, after every rain, from the middle of July to the first of September. In these plants I found only eight borer cavities, *all of which were farther from the base of the plants than the Paris green water was applied.* In eight hills of the same variety I placed cobs dipped in coal-tar, five in each hill; following out the suggestion that you gave me, I re-dipped the cobs from time to time during July and August. In these plants I found but three borer cavities.

" In sixteen hills of the Perfect Gem squash, where nothing was applied, I found fifteen borer cavities. In eight hills of the same, where I used the soap emulsion on the stems, in the same manner as I used the Paris green mixture in the case noted above, I found but two borer cavities.

" I think that these results give hope that we may very largely diminish, if not entirely remedy the damage wrought by this insect, by a thorough application of insecticides to the stems, or the use of counterodorants."

In the above experiments, the application of the soap emulsion seems to have given the best results, provided that the " Perfect Gem " squash is as subject to attack as the " Hubbard," as eight hills gave but two burrows.   Next to this comes the coal-tar employed as a counterodorant, showing in the same number of hills but three burrows ; and, third, the Paris green water with eight burrows.

The average protection shown in the above experiments (not including that with the coppcras water) is eighty-one per cent — a highly gratifying result.

Had the applications been commenced about the first of July, instead of the middle of the month, it is probable that the results would have been still more satisfactory. The moth, in the State of New York, may commence the deposit of its eggs during the first week of July.   The coal-tar would act by preventing the deposit, while the Paris green water and the soap emulsion operate by killing the young larva upon its hatching from the egg and beginning to eat into the vine.

In a subsequent communication, Mr. Goff, in reply to inquiries, contributes these items relating to the life-history of the species :

"Strange to say, I found but two live larvæ ; in almost every case they had left the vines for transformation. * * * * * * I found no pupæ within the vines.   In one or two cases the caterpillar seemed to have died ; at least, I found in the cavity a mass of white matter. * * * * * * I noticed that the borers are often found in the stem at a greater distance from the base than I had supposed — often four and even six feet from the base. Strange to say, however, in the ' Perfect Gem ' squash I found the borer only close to the base, in no case, I believe, in a *branch*, but always in the main stems. * * * * * * The ' Perfect Gem ' seems less liable to attack from the borer than the ' Hubbard.' I had noticed this fact in previous seasons."

Orgyia leucostigma (Sm.–Abb.).

The White-Marked Tussock Moth.

(Ord. LEPIDOPTERA: Fam. BOMBYCIDÆ.)

SMITH-ABBOTT: Nat. Hist. Lep. Ins. Georgia, 1797, ii, p. 157, pl. 79 (figs. larva, pupa, ♂ & ♀ ins.).

HARRIS: Rept. Ins. Mass., 1841, p. 262; Treat. Ins. N. Eng., 1852, pp. 282–284; Ins. Inj. Veg., 1862, pp. 366–368, pl. 7, figs. 1–5 (natural history, etc.).

FITCH: in Trans. N. Y. St. Agricul. Soc. for 1855, xv, 1856, pp. 441–450; 1st–2d Rept. Ins. N. Y., 1856, pp. 209–218 (detailed account); 3d–5th Repts., 1859, p. 20 (brief notice); 6th–9th Repts., 1865, p. 109 (increased injuries and remedies).

RILEY: 1st Rept. Ins. Mo., 1869, pp. 144–147, figs 81–83.

MORRIS: Synop. Lep. N. A., 1862, p. 249 (descr. of larva and imago).

PACKARD: in Proc. Ent. Soc. Phila., 1864, iii, p. 331.

LE BARON: 1st Rept. Ins. Ill., 1871, pp. 13–17.

SAUNDERS: in Canad. Entomol., iii, 1871, p. 14, f. 10 (eggs and larva descr.); in
  Rept. Ent. Soc. Ont., for 1874, p. 19, figs. 14–16 (general notice); Ins.
  Inj. Fruits, 1883, pp. 57–60, figs. 50–53 (the several stages and remedies).
BETHUNE: in 1st Rept. Ent. Soc. Ont., 1871, pp. 82–3, figs. 23, 24 (brief notice); in
  2d Rept. Id., 1872, p. 14 (abundance and injuries).
FRENCH: in Thomas' Rept. Ins. Ill. (vii), 1878, pp. 185–186, figs. 36, 37 (general
  notice).
FORBES: 12th Rept. Ins. Ill., 1883, pp. 100–1, fig. 20 (brief notice).
COLEMAN: in Psyche, ii, 1882, pp. 164–166 (larval colorational differences).
LINTNER: in Albany Evening Journal for June 25, 26 and 28, 1883; in 37th Rept.
  N. Y. St. Mus. Nat. Hist., for 1883, pp. 50–52 (larvæ girdling twigs).

In studying the history of our insect foes it not unfrequently occurs
that a species which had long been known to entomologists, without
having assumed such injurious habits as to give it place among our
serious pests, suddenly increases to such an extent and so extends the
range of its food-plants, as to become a conceded and permanent public
nuisance.

Such an one is the *Orgyia leucostigma*, an insect broadly distributed
over the United States, from Maine to Georgia and westward to the
Rocky mountains. For many years it was known only as an occasional
depredator upon apple trees and rose bushes. In the year 1828, according to Dr. Harris, the apple trees in portions of New England were infested with multitudes of the caterpillars, and subsequently, in the summers of 1848, 1849 and 1850, they extended their attack to many other
trees, nearly stripping the leaves from the horse chestnuts in Boston.

It appears for a long time to have had but local distribution. Dr. Fitch,
writing of it in 1856, had never known it to be sufficiently multiplied in
the vicinity of his residence, Salem, Washington county, N. Y., to merit
any attention on account of its depredations. He had never met with
a half-dozen of the caterpillars in any one year, perhaps owing to the
locality being near the northern extreme of their geographical range,
until in the summer of 1855, when they were noticed as being unusually
common.

As early as the year 1870 it had become abundant in Albany and
firmly established. It has annually reappeared, with perhaps three or
four exceptions, and made great havoc upon the foliage of both fruit and
shade trees. One of these exceptional years was in 1874, when but few
of the larvæ were seen, and another in 1884, when these notes are being
revised for publication. In several other of our larger cities this insect
also became a public pest. In Philadelphia they were very abundant
in 1873, as appears in an article in the *Popular Science Monthly*, vol.
iv, page 381, entitled, "The Caterpillar Nuisance in Philadelphia," in
which a former pest [*Ennomos subsignaria*], one of the measuring-worms

exterminated by the English sparrow, is replaced by the *Orgyia*, which the sparrows will not attack.

As will be seen from the literature of this species as given above, it has already been treated of by most of our writers on economic entomology. It has also been noticed in many other minor publications not cited, and the figures originally given of it in its different stages often reproduced. But so serious are its injuries to fruit and shade trees and various ornamental shrubs, and so general the ignorance of the best means of relief therefrom, that the present notice of it seems desirable.

As a knowledge of the several stages of insect life and other particulars which enter into a life-history are of great importance in all efforts to control insect depredations, they will first be considered.

The Caterpillar.

While the perfect insect is unusually sombre in color and without the slightest ornamental feature, unless it be its prettily pectinated antennæ, the caterpillar, as if by way of compensation, is strikingly beautiful, not only to naturalists, but to all who will take the trouble to give it an examination. Dr. Fitch has written appreciatively of its beauty, as follows:

" The term ' caterpillar ' is applied to a worm which is clothed with hairs ; and we commonly associate this term with something which is ugly and repulsive in its appearance. But many caterpillars are far from meriting this prejudice, being, in reality, objects of much beauty. This is eminently the case with one which may frequently be seen in the month of July upon apple trees and also in our yards upon rose bushes. We cultivate the rose for ornament, and nature, as if to further our designs, places upon the leaves this neat, little, prim caterpillar, which is a more delicate, elegant object than the handsomest rose that ever grew. I well remember the first time I ever noticed one of these caterpillars. It was in the hay field, in my boyhood. One of the laborers, who had little taste for any of the beauties of nature — a man of that class of whom the poet sings,

> ' A primrose by the river's brim,
> A yellow primrose was to him,
> And it was nothing more ' ——

in stooping for a handful of grass to wipe off his scythe, had his attention arrested by one of these caterpillars. Taking up the leaf on which it was standing, he was for several moments absorbed in contemplating its bright colors and the artistic arrangement of its elegant plumes. Then as he was laying it down, he said to himself, ' That is the prettiest thing I ever saw ! ' Let us not murmur if the leaves of our rose bushes are

somewhat gnawed and eroded, when they hereby produce for our admiration objects far more beautiful than we look for them to yield."

The caterpillar is a slender creature, measuring when full-grown from
about three-fourths of an inch to an inch and a quarter in length.   Its
color is a cream yellow, with a broad velvety black stripe upon its back
and a broader brown or blackish one upon each side.   Nearly covering

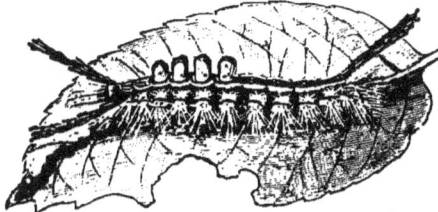

its sides are tufts of pale
yellow hairs, which radiate
from small yellow tubercles
arranged in two rows : from
the upper of these rows a few
longer black hairs are given
out.   Upon the back, on its
anterior half, are four erect,
short, thick, even, brush-like

FIG. 7.—Larva of the White-Marked Tussock moth, ORGYIA
LEUCOSTIGMA.

tufts of hairs, white or creamy-yellow, placed on the fourth, fifth, sixth
and seventh rings of the body.   The contrast between the occasional
bright yellow of these tufts and the white has been so marked as
to give rise to the suspicion that they might indicate sex, but the careful
observations of Mr. Coleman (*loc. cit.*) leave no ground for such belief.
The same larva has, at different times, shown yellow tufts and white
ones, changing from yellow to white and from white to yellow, in-
dependently of molting.   From the sides of the front ring project
forward in a broad **V** shape, two pencils of black, bearded hairs, tufted
at the end and of unequal length, giving them a jointed appearance,
the longest measuring a half-inch in length.   On the penultimate ring
(11th) is a similar pencil of brown and black hairs, which are only tufted
at the end of the pencil.   The head is coral-red, as are also two little
knobs on the back within the black stripe, on the ninth and tenth rings.
Fig. 7 shows the caterpillar.

The Cocoon.

Next following the caterpillar stage, the insect is presented to us
under the guise of a cocoon, attached usually to the trunk or branches
of the tree upon which it had fed.   Frequently, however, it wanders from
its food-plant and seeks more hidden and sheltered locations, as under
window-sills, beneath copings, caps of fence-posts, etc.   The cocoon is
about an inch in length, broadly flattened beneath, of an elongate-oval
outline and of a depth barely sufficient to contain the pupa.   It is very
slightly woven, for, although it is double, consisting of an outer and an
inner web, yet, through both, the larva, near its transformation, or the
pupa, may often be seen.   Woven into the outer envelope are notice-

able the long, black pencil-hairs of the caterpillar and some of the shorter yellow ones.

The Pupa.

Upon tearing open the cocoons, the pupæ found therein (the third stage in the insect's transformation) will be observed, if a number are examined, to vary greatly in size — some being less than a half-inch in length, and others exceeding seven-tenths of an inch — these last surpassing the former four or five times in bulk: the smaller ones produce male moths, and the larger, females. They are rounded in front and in their entire outline, except at the hinder end, where they terminate in a short spinous tip armed with bristly hooklets by which they fasten themselves to the threads of the cocoons. The colors are whitish and brown approaching black, particularly upon the back. On the front, back and sides are numerous short white hairs—a quite exceptional feature in the almost universal smoothness of the pupal form.

Fig. 8.—The White-marked Tussock Moth: c, the female pupa; d, the male pupa.

A prominent and characteristic garniture of the female pupa, shown at c, in Fig. 8, is a pale, oval or square spot composed of minute bladder-like bodies or scales on each of the principal rings (the thoracic ones) back of the head. In the male, d, in the figure the broad antennæ-cases, with their transverse markings denoting the pectinations of the organs, are conspicuously folded upon the breast beween the wing-cases.

The Perfect Insect.

The male moth, although far superior to the other sex in outward appearance, is not an object that would readily arrest the attention — its plain colors being in marked contrast with those worn in its caterpillar stage. Its size is rather below the average of moths and is seemingly small for the larva producing it, as it measures but little more than an inch across its expanded wings. The wings are broad and rounded in outline and of a smoky-brown color. The front pair have a gray patch on the middle of the anterior border, and are crossed by three or four blackish curved lines, of which the outermost one is quite angulated; near the outer margin is a gray band, preceded near the tip by some elongate black spots, and near the inner angle by a white spot, often comma-like in form. It is from this last that its specific name of *leucostigma* (white-mark) is derived. The hind wings are also broad, smoky-brown, without any markings, but with a darker border. Both pairs are fringed with projecting scales.

Fig. 9.—The White-marked Tussock Moth, ORGYIA LEUCOSTIGMA — position at rest; the male

The body is brown, quite slender, small in proportion to the spread of wings, and bears a small black tuft of upright hairs near its base.

The antennæ are beautiful objects — the redeeming feature of the unusual plainness. " They are about one-third of the length of the wings, gray, with a double row of dark-brown branches resembling the teeth of a comb. Each branch has a row of very fine hairs, like eye-lashes, along each side, and at its tip three bristles, one of which is much longer and directed inward toward the head " (Fitch).

When at rest, the moth has its long, hairy front legs stretched out divergingly in front of its head, with the antennæ usually extended at right angles with its body, as shown in the figure. With the recollection of this unusual attitude in connection with the features above indicated, the moth can hardly fail of being recognized during its season of prevalence — toward the last of July and early in August.

The female is of a very different appearance, being apparently wingless, having only the rudiments of wings in the form of a little pad on each side, and its ash-gray body distended with its burden of eggs when it creeps out of the cocoon, to a breadth of a quarter of an inch in a length of half an inch. Its antennæ are not branched, and it scarcely possesses any feature which would lead the ordinary observer to associate it with its mate or to regard it as a moth.

The Egg.

The eggs — the last phase as we have traced the progress, but properly, the first in the series of the four insect stages — are deposited in mass upon the surface of the cocoon, as will be hereafter described. The egg is round, with a strong, thick, cream-colored shell, and in size is about three-hundredths of an inch in diameter. It has a slight depression apically, marking the point at which the caterpillar eats an opening for its escape.

Natural History.

The life-history may next be traced in its progress through the several stages above described. The eggs are hatched at about the middle of May, varying with the season, and, perhaps, from their different exposures, for in the same locality some of the deposits have been from three to four weeks later than others in disclosing their larvæ. As has been followed by Prof. Riley; the stages of growth observed were as follows :

The first molting of the caterpillars occurred seven days after escaping from the egg. At intervals of six days each, the second and third moltings took place. Six days thereafter some of the larvæ had attained their growth, ceased feeding, and commenced to spin their cocoons.

10

Others, at about this time, again molted (for the fourth time), and not until after six more days did they commence inclosing themselves in cocoons. When the moths emerged, it was discovered that males were produced from the larvæ which had molted but three times, and females from the remainder.

This prolonged larval period and additional molting in the female is a very interesting scientific fact. Its occurrence in other species, as observed by me, has been noted in my first report (page 98), and it is believed that it will be hereafter discovered to be the rule in many other of the *Bombycidæ*. The bearing of this fact upon the mooted question of the comparative degree of development and rank of the sexes is worthy of consideration.

To summarize the above statement of the successive larval stages — the entire period is from twenty-five to thirty-one days — about a month. It is during this brief period that the voracious appetite, necessitated by rapid growth, causes the ravages upon our trees and shrubs of which we have so great cause to complain.

Within the shelter of the cocoon, the larval skin is cast off for the last time, and the insect, as early as the latter part of June, enters upon its pupal state. In this condition it quietly reposes for a period of from ten to fifteen days,* when the moth within the pupal shell, having by this time become fully developed in all its organs except in the extension of its wings (in the male), bursts its walls, and emerges to the light of day. The male moth — first observed by me the present season on the 9th of July— crawls to some suitable spot near to the vacated cocoon, when, within a half-hour, his contracted wings, as drawn from their pupal sheathes, attain their full expansion, and he is in readiness for flight. As soon as "evening shades prevail," he seeks his mate, and with the end of his existence attained, a few days serve to end his career.

The female moth hardly surpasses the "Bag-worm," *Thyriodopteryx ephemeræformis*, as a traveler, for while that species never emerges from her cocoon, this only works its way outward through the loosely-spun threads, and takes position upon its surface. From a power of attraction possessed by her in an eminent degree,† and shared by many of

---

*Cocoons spun on July 5 were found with egg-deposits upon them on July 16. Dr. Fitch records that in several instances he had known the moth to emerge thirteen days after spinning the cocoon.

†A female imago of this species had emerged August 4, within its breeding-cage, standing in a large apartment about ten feet from an open door. At dark (half-past seven o'clock), males commenced to fly into the room, and precipitate themselves against the gauze front of the cage, moving in every direction over its surface, with legs, wings and antennæ in rapid motion, in a persistent effort to force an entrance into the cage. Several

the *Bombycidæ*, it rarely, if ever, happens that she is not prepared to deposit her burden of fertilized eggs on the following day; and as if aware that her size and exposed position was an invitation to her many enemies, she enters upon the duty without delay. A connected thread of eggs is extruded from her body and attached in mass to the surface of the cocoon. As they emerge from the abdomen, they are accompanied with a viscid secretion which cements them firmly together; and when the entire number — two hundred or more — are deposited, they are thickly covered over by the same frothy substance, which soon hardening, serves as an admirable protection from unfavorable weather, from other insects, many birds and other depredators. This accomplished, the shrunken body soon drops from the cocoon to the ground, and the life-cycle of the individual is completed.

A second brood. — In the Southern States there are two annual broods of the insect, of which the moths of the first appear in May and June, and those of the second in September and October. Even so far north as Philadelphia, I have observed on the 8th of September (after a long term of unusually hot weather), an abundant second brood, at which time the mature larvæ abounded upon the tree-trunks, together with cocoons recently spun and in construction, females just emerged, and egg-deposits upon the cocoons.

In New York, a second brood occurs occasionally, at least; and perhaps annually in limited numbers, yet it is never so abundant as to be injuriously noticeable. I have seen the moth as late as October 6th.

Hibernation. — The hibernation of this species is in the egg stage — the exception with the lepidoptera, where it is commonly in the pupal stage, often in the form of larvæ, and occasionally as perfect insects. The greatest cold of our northern winters — from 30° to 40° below zero, of Fahr. — has failed to kill them. How it is possible for them to survive such temperatures is a mystery to us. It is not from the protection afforded them by the material in which they are encased, for

FIG. 10.—Egg-belt of HEMILEUCA MAIA.

the eggs of another Bombycid—*Hemileuca Maia*—survive the winter equally well, although they are entirely without any covering and are exposed in belt-like manner upon the twigs of oaks,

attempted to enter through the small crevice left by the imperfectly fitting door at the rear of the cage. Three or four moths were often on the gauze at the same time, whence they could be plucked with the thumb and finger. During the hour that this exhibition continued, forty moths were taken and pinned, from at least a hundred that entered the room. *Twenty-sixth Report on the New York State Museum of Natural History*, 1874, p. 148; id., *Entomological Contributions, No. III*, p. 148.

as represented in Fig. 10.* The eggs of butterflies having still thinner shells than the above, have been exposed for hours to freezing mixtures of —22° Fahr., without injury to them. It is probable that the eggs of most of our insects are in reality not frozen under the greatest cold to which they would be subjected under natural exposures. Some eggs of *O. leucostigma* which I had taken separately from their covering and left exposed during a portion of the winter, were examined by me under a temperature of —18° Fahr., and were found to be in their natural fluid condition.

Food-Plants.

An important inquiry, whenever a noxious insect presents itself for investigation, is its range of food, to serve as a guide to the probable extent of its depredations, and the possible means of checking its spread. In the year 1861, when the citizens of Brooklyn were greatly excited over the increase of a "measuring-worm,"† which had become an unendurable nuisance from the myriads that hung suspended by their silken threads from the branches of the shade-trees of the principal streets, the following resolution was discussed by the common council:

"*Resolved*, That the owners of property having linden trees on the streets are hereby ordered to remove them within ten days after the passage of this ordinance, and failing to do so, shall be liable to a fine of five dollars for each tree left upon the streets, after that date. The street commissioner is hereby ordered to remove all trees of that species that may be left on the streets after that date."

The resolution having been referred to a committee, failed to be reported favorably and to become a law, for the reason that upon scientific examination it was found, that while the caterpillar was more abundant upon linden trees, it also occurred and fed upon the elm, weeping-willow, silver-leaved poplar, balm-of-Gilead, maple, horse-chestnut, some of the fruit-trees, shrubs and herbage.

Occasionally an insect seems to be so extremely particular in its food as to confine itself to a single species, but more commonly it is found feeding upon the several species of the same genus. Others extend their range to the several genera of the same family, while others still, known to us as polyphagous species, attack plants of very dissimilar characters, comprised under several orders. In illustration of the last, we may cite the habits of two species of well-known *Bombycidæ*, viz.: *Samia*

---

* *Twenty-third Report on the N. Y. State Cabinet of Natural History*, 1872, pp. 137–141; *Entomolog. Contrib.* [*No. I*], 1872, pp. 5–9.

†The larva of the snow-white linden-moth, *Ennomos subsignaria* Hübn.

Cecropia (Linn.) and *Telea Polyphemus* (Linn.). The former is recorded as feeding upon forty-nine species of plants,* belonging to nine-
teen genera, and to nine natural orders — the orders being *Tiliaceæ,
Aceraceæ, Rosaceæ, Saxifragaceæ, Caprifoliaceæ, Urticaceæ, Cupili-
feræ, Betulaceæ* and *Salicaceæ :* the latter, *Polyphemus,* upon the same
number of species, in sixteen genera, and in eight orders. (Wm. Bro-
die, in *Papilio,* ii, 1882, pp. 32–3, 58–60.)

The Orgyia caterpillar has not been favored with so careful a compilation of its food-plants ; the following, presumably the more common
ones, are recorded : Linden (*Tilia Americana*), horse-chestnut (*Æscu-
lus hippocastanum*), locust, rose, plum, cherry, apple, pear, elm, butter-
nut (*Juglans cinerea*), black-walnut (*Juglans nigra*), hickory (*Carya
alba*), oak, spruce, fir, and larch.   I noticed it last summer, in September, seriously injuring the foliage of some castor-oil plants, *Ricinus
communis.*

Depredations.

On the horse-chestnut. — The horse-chestnut, which had steadily been
growing in popular favor as a shade-tree, from the almost entire immunity which its foliage has enjoyed from insect attack,† has of late years
been apparently chosen by the Orgyia for its favorite food-plant.   It
might be thought that a leafage so dense and luxuriant would afford more
than the required amount of food for any number of insect depredators
that could assemble upon it; but in seasons favorable to the multiplication of the insect, the largest trees in the city of Albany have shown
the pitiable spectacle of a foliage entirely consumed — only the ribs of
the leaves remaining.   Not only were the trees left useless for shade, but
their lives were endangered, for, although under favoring conditions of
moisture and not excessive temperature, a second feeble leafage would
sustain their existence, yet a repetition of the attack the following year
could hardly fail of proving fatal.   Hosts of greedy caterpillars could
be seen deserting their exhausted feeding grounds and descending the
trunk in search of the food still needed by them.

On the elm.— The elm is also injured to a degree not much less than
the horse-chestnut.   The immense size of most of the white elms
(*Ulmus Americanus*) planted for shade, offers, in their long limbs and
spreading branches, more abundant food, and in a measure diffuses the
ravages that they sustain, yet frequently a large proportion of their

---

* Not such as the cocoon and larva may have been found upon, "but held to be such as
a perfect female *Cecropia,* at liberty, will select as food for her young."

† Besides this species, I can recall but two other caterpillars that feed upon it, viz.,
Acronycta Americana Harris, and *Ennomos subsignaria* (Hübn.), and these, usually, to
quite a limited extent.

leaves have been skeletonized, or so riddled that they wither, dry, and drop to the pavement.

On fruit-trees.— Dr. Le Baron, has recorded that "on the second day of September my attention was called to an orchard a few miles from my residence [at Geneva, Illinois], in which all the trees in one corner of the inclosure, to the number of fifteen or more, had been entirely stripped of their foliage by these caterpillars, whilst they were at the same time well loaded with fruit."

Dr. Fitch states that in 1863, in Albany, they completely stripped large plum trees of their leaves, and rose-bushes were similarly defoliated. (*6th to 9th Repts. Ins. N. York*, 1865, p. 199.)

On Cape Cod, Mass., in 1884, when the larvæ generally were less abundant than usual, they are reported by Mr. J. B. Smith as stripping the leaves everywhere. (*Canad. Entomol.*, xvi, p. 183.)

Some plum trees that were under my almost daily observation were nearly stripped of their leaves before the attack was noticed.

In Central Park, New York.— During the summer of 1883 the insect was very prevalent in many portions of the State of New York and in adjoining States. In New York city the extensive defoliation of the shade-trees in midsummer was felt as a public calamity. In Central Park the ravages were particularly noticeable. When they had reached a point that threatened the destruction of many of the infested trees, the park commissioners were compelled to take action to arrest, if possible, the attack. The collection and destruction of the eggs was ordered. The work was undertaken under the direction of Mr. E. B. Southwick, in charge of the inspection of trees and shrubs, and, as the result, as he has informed me, between the 13th of August and the 16th of October, the four men engaged in the work, armed with hooks and scrapers of various kinds, steel brushes and brooms, went over the entire park and collected *thirteen bushels of the Orgyia eggs and cocoons* and burned them.

Parasites.

Upon one occasion Dr. Le Baron found the cocoons of this species, which were very abundant upon some apple-trees, extensively infested by the pupæ of a Tachina fly, which was named and described by him as *Tachina orgyiæ* (*1st Rept. Ins. Ill.*, p. 16). Very few of the cocoons examined were free from the Tachina parasite ; two, three and sometimes four were contained in them. Of the myriads of the cocoons observed, it appeared that scarcely one out of a hundred had escaped the fatal visitation of this parasite.

From one of the caterpillars Dr. Fitch bred a number of minute parasites belonging to the family of *Chalcididæ*, which were described by him under the name of *Trichogramma? Orgyiæ*. Another species, resembling this so closely as to have been named by him the " brother parasite," *Trichogramma fraterna*, was found by him under conditions that led him to believe that it was also parasitic on the *Orgyia*. These species have been subsequently referred to the genus *Tetrastichus* by Mr. Howard, who is of the opinion that they will prove to be not parasites upon the *Orgyia leucostigma* larva, but upon a species of *Pteromalus* which is parasitic upon it. This *Pteromalus* has been reared by me in large numbers, but, in the uncertainty that attends the numerous described species of this extensive genus, it remains at the present undetermined.

I have also obtained from the eggs of the *Orgyia* a minute parasite, which has been generically determined by Mr. Howard as *Telenomus*.

A species of *Hemiteles* is a very common parasite upon the larva in the vicinity of Albany, and in some seasons it has proved to be the principal instrument in arresting its multiplication.

There is also in my collection a Tachinid fly obtained from an *Orgyia* in 1879, which differs from the species described by Dr. Le Baron, in that it has two large reddish spots on each side of its abdomen, and that its white tegulæ, instead of being translucent, are quite opaque. The *Tachina orgyiæ* is not at all uncommon with us.

The Great Increase in its Ravages.

That this insect has very largely increased in numbers and in destructiveness during the last twenty years is evident to all who have followed its history during that time, or to those who have but recently been drawn to notice it as a new depredator. To what cause, it may properly be asked, is this great increase owing?

An unusual abundance of a particular species of insect for a single year is a phenomenon of common occurrence, which can easily be accounted for from the operation of well-known natural causes. A remarkable abundance of a species for a term of years is not infrequent, but soon it relapses into its former paucity and harmlessness before the attack of hosts of parasitic foes, which its great increase invites.

Not so, however, with the *Orgyia*. Its conspicuous ornamentation and its assemblage in large numbers render it an attractive object and easily to be discovered by its enemies, and already, as stated, seven different species of insects are known to prey upon it. Scores of the smaller ones (*Chalcididæ*), are often feeding in company within the body of a single caterpillar, and it sometimes seems as if almost every cocoon gave forth the parasite instead of the moth. And yet the destructive caterpillar abounds, and continues to extend its range and to increase its injuries.

That the present attitude of this insect is so different from that which it formerly held, and which it had assumed, doubtless, many years ago* in conformity to a prevailing law of adjustment in nature, by which each creature eventually becomes fitted to the place it holds — cannot but be the result of some disturbing element operating upon it. Nor can there be any reasonable doubt of what this interference has been.

The English Sparrow the Cause of the Increase.

The extraordinary increase of the *Orgyia leucostigma* is owing to the introduction and multiplication of the English sparrow, *Pyrgita domestica*.

This may seem a strange statement to many, in consideration of the fact that the sparrow was imported from Europe for the express purpose of abating the " caterpillar nuisance " in New York and some of the New England cities.

In the year 1866, two hundred sparrows were brought over from England, and released in Union Park, New York. In 1867, forty pairs were imported and set free in New Haven.

It is claimed in behalf of the sparrows that they accomplished the object of their introduction, and that within two years, they arrested the " measuring-worm plague," which had for years prevailed in the streets and parks of those two cities. They may have done so.† If it could be shown that they rendered this service — perhaps the only good that they have done — it would be insignificant in comparison with the harm they have subsequently caused — even if compared alone with the protection given by them to the *Orgyia*.

In substantiation of this charge, we present some notes of personal observations made in Albany, during the year 1872, which might doubtless have been supported by similar observations in many other localities.

From the rapidity of multiplication natural to the English sparrow, and from the encouragement extended to them by our citizens in providing cages for their shelter and food during the winter, they have become quite numerous. In several quarters of the city where they have

---

*It is a native species, named and described in 1797.

†A correspondent of the *Rural New Yorker*, in the issue for January 23, 1875, says: " At the very time of their introduction into New York City and Brooklyn, a small ichneumon fly had already lessened the number of span-worms which were so disagreeably abundant in these cities, and it is very probable that the insects would have disappeared without the aid of the birds."

established themselves, their little flocks are continually before the eye, and their unceasing chatter constantly in the ear, almost to the exclusion of other sounds.

The increase of the *Orgyia leucostigma* commenced and has continued to progress with that of the sparrow.

A remark made to me that the caterpillars had been observed to be very numerous in localities where the sparrows also abounded, induced me to undertake to verify or disprove the idea that had suggested itself to me, that the sparrow afforded actual protection to the caterpillars, and promoted their increase.

In a locality in the city (intersection of Broadway and Spencer streets), which I had traversed daily during the preceding year, I had been interested in watching the habits of a large company of sparrows which had established themselves in quarters evidently in every way suited to their taste and wants, among the vines and leaves of a large woodbine (*Ampelopsis quinquefolia*) which covered with a dense matting nearly the entire side of a large dwelling. Here I had observed a greater number of the sparrows than elsewhere in the city: they were still local, and far from being generally distributed.

Upon visiting this locality for the purpose above mentioned, I found upon the other side of the building and on an adjoining one, three other large woodbines not before noticed by me — making five in all. On a tall pole standing between the two buildings, a very large sparrow house, with many compartments, had been erected, and many smaller ones had been placed among the branches of the trees. The woodbines seemed alive with the sparrows : hundreds were issuing from them and dropping down to their favorite stercoraceous repasts in the streets, and the air was vocal with their chattering. It was a rare bird-exhibition. Here certainly was a test-case of the "insectivorous nature" of the sparrow.

On the sidewalk in front of the two buildings, two large, spreading elms (*Ulmus Americanus*) standing between some maples, showed *every leaf eaten from them*, disclosing the nesting-boxes among their branches, and their trunks and limbs dotted thickly, or clustered with the easily-recognized egg-bearing cocoons of the Orgyia. Hundreds of immature caterpillars were traveling over the trees, fences, and the walls adjoining. No better evidence of the almost perfect immunity afforded to the caterpillars from their enemies — whether birds or insects — by the presence of the sparrows, could possibly be given.

A portion of Broadway, between Clinton avenue and the Central railroad crossing, was also known to abound in the sparrows, the citizens

11

resident there having fed them most generously, not only during the winter season, but also in the summer months. Nesting-boxes had been placed for them in most of the trees. Here the trees presented a pitiable sight. Many of the elms and horse-chestnuts were entirely stripped of their foliage — the naked ribs of the leaves of the latter seemed ghastly in their suggestion of fleshless fingers. Nowhere else in the city had I seen such ravages.

Passing thence to Pearl and State streets, the same association of sparrows, caterpillars, and their destructive work was seen. Clinton square, where the sparrows had, in their introduction into the city, been specially taken under the care and protection of the residents on the east side of the park, afforded another excellent " test." It was evident that the sparrows were in full appreciation of their privileges, from the almost incredible number sporting among the trees. Their protégés were also in full force. Caterpillars and their cocoons met the eye everywhere, while hanging from the rails and caps of the iron fence surrounding the park were the dead and decomposing bodies of caterpillars, killed by the recent heavy rains (often so fatal to insect larvæ), in such numbers that they tainted the air in their vicinity.

It seems unnecessary to extend this record farther than to add that in other sections of the city, observations made, were in accord with the above.

How the Sparrows protect the Caterpillars.

That the sparrows decline to eat the Orgyia caterpillars is not a charge against them : they *could* *not* eat them with impunity; the diet would doubtless prove fatal to them. The charge to which they are amenable is this: By the force of numbers united to a notoriously pugnacious disposition, they drive away the few birds that would feed upon them. Of these we know but four species, viz.: The robin (*Merula migratoria*), the Baltimore oriole* (*Icterus galbula*), the black-billed cuckoo (*Coccygus erythropthalmus*), and the yellow-billed cuckoo (*Coccygus Americanus*). The above species seem, in the ordering of nature, to have been assigned to us for protection from an undue multiplication of a large number of hairy caterpillars of injurious habits. While the naked ones are a tempting morsel to most of our insectivorous birds, they instinctively reject the others, the hairs of which would fill their stomachs with a mass of irritating and indigestible material. As we often discover in the animal kingdom special provisions for the accomplishment of certain purposes — as for example, the crossed mandibles of the cross-bill

* This bird has been seen with its head thrust into the web-nest of the tent-caterpillar, eagerly devouring its occupants.

(*Loxia leucoptera*), fitted for tearing open the cones of hemlock and pines for extracting their seeds for food — we may presume that the above-mentioned birds are enabled, through some peculiar structural feature, to appropriate the food that others are compelled to reject. One of them, the yellow-billed cuckoo, is known to shave off the hairs of the *Orgyia leucostigma* caterpillar before swallowing it. The following account of the operation is from Dr. Le Baron, former State Entomologist of Illinois: " My attention was attracted to a cuckoo regaling himself upon these caterpillars which were infesting, in considerable numbers, a larch growing near the house. My curiosity was excited by seeing a little cloud of hair floating down upon the air from the place where the bird was standing. Upon approaching a little nearer, I could see that he seized the worm by one extremity, and drawing it gradually into his mouth, shaved off, as he did so, with the sharp edges of his bill the hairy coating of the caterpillar and scattered it upon the wind" (*First Report on the Insects of Illinois*, 1871, p. 17).

Preventives and Remedies.

A relentless war upon the English sparrows.— If the protection of the Orgyia by the sparrow has been shown, as we believe, beyond question, it follows that, with the abatement of the sparrow plague, the ravages of the caterpillar would be arrested, or very greatly reduced. In view of the magnitude of these injuries to our shade and fruit trees, I do not hesitate to recommend, as a preventive measure, the removal, from among us, of the English sparrow. It would also serve to diminish the losses annually sustained in our orchards, forests and gardens, from the following well-known noxious species : The apple-tree tent-caterpillar (*Clisiocampa Americana*), the forest tent-caterpillar (*Clisiocampa sylvatica*), the fall web-caterpillar (*Hyphantria textor*),* the yellow-necked apple-tree caterpillar (*Datana ministra*), the yellow woolly-bear (*Spilosoma virginica*), and many others of the kind.

As means toward the desirable removal of this intruding protector of injurious insects, the "*ineligibility*" of which in this country has been clearly shown by high ornithological authority,† the following may be suggested: Removal of all nesting-boxes; destruction of their nests;‡ depriving them of shelter and nesting beneath the cornices of buildings

---

* For its increasing abundance in Washington, D. C., see *American Naturalist*, Sept., 1881, xv, p. 747.

† See Dr. Coues on " The Ineligibility of the European House Sparrow [*Passer domesticus*] in America." *American Naturalist*, August, 1878, xii, pp. 499–505.

‡ Under the existing laws of the State of New York, this is only permissible under certain circumstances. It is hoped that all restriction will be removed by the present Legislature, through a proposed amendment to our game laws.

by wiring over all available openings ; using them for food, they being already an article of sale in markets, and said to make an excellent pie ; employing the most efficient means to prevent their living in grain-fields, where their intrusion forfeits their lives * (not protected by law). Still another means has been reported to me as having been resorted to under the provocation of great annoyance and large losses — feeding with food prepared with some anæsthetic, *permanent* in its effects.

Destruction of the Orgyia eggs.— An effectual means of prevention of the Orgyia injuries, so far as it can be resorted to, is the destruction of its eggs. This may be accomplished, to a great extent, by a moderate expenditure of labor. The egg-deposits are conspicuous objects, and at once attract the eye looking for them, until, after long exposure, they become weather-worn and lose their original whiteness. Even then they are not difficult to find, as the cocoons upon which they are placed often occur in patches of a dozen or more. A favorite location for them is where a limb has been cut from a tree and a convenient angle afforded by the overlapping growth of the sap-wood. As many as fifty egg-bearing cocoons have been gathered from a single tree-trunk while standing on the ground, the eggs of which represented twelve or fifteen thousand caterpillars.

The month of August would be the most favorable time for collecting the eggs from the trunks and larger branches of trees, they being more conspicuous at that time than later. Those out of arm's reach could be removed by poles or scrapers prepared for the purpose.

---

* The sparrow has increased greatly during the last ten years [in England]; great packs of them swoop down on the wheat-fields, *destroying more than they consume,* spilling it over the ground. Every piece of wheat that I saw this year has had more corn [wheat] destroyed by sparrows than would pay one rent at least. *   *   * If there are as many sparrows in other parts of the country as there are in Lincolnshire, most certainly *one million pounds sterling* would not repay the occupiers of the land for the yearly loss sustained by the depredations of this most quarrelsome pest. *   *   * They prevent the increase of swallows, and have literally driven all our soft-billed insect-eating birds from our gardens and orchards. The fly-catcher has gone ; also the tree-creeper, the peep, the minor warblers, most of which lived on the eggs of moths and butterflies. *
*   * They were found to be feeding upon turnip seed   *   *   *   to be eating red clover seed   *   *   *   for two weeks they ate buds of fruit trees.
*   * In August, just when the grain begins to ripen, they assemble in vast flocks, and soon commit sad havoc in fields of wheat, oats and barley.   *   *   *
A field of wheat near Isleworth was so utterly ruined by legions of the sparrows which swarm amongst the neighboring villas, that it was left uncut. (*Report of Observations of Injurious Insects,* by Eleanor A. Ormerod, 1884, pp. 40-42.)

Such will probably be the habits of the sparrow when it shall have spread over the agricultural districts of the United States (at present preferring the larger cities), unless a war of extermination be soon commenced, and continued under the encouragement of liberal bounties offered for their heads.

In England, at the present time, farmers are resorting to poison as an easy way to reduce their numbers, and Farmers' Clubs are paying from 3d to 6d per dozen for old or young sparrows or their eggs.

In a communication to the *Albany Evening Journal* of August 13, 1881, the following recommendations were made by me :

1. That each householder charge himself with the duty of the collection and destruction of these egg-masses upon his own premises. The children of a family will prove ready collectors, especially if encouraged by competition or the offer of a small reward.

2. That the authorities in charge of our public grounds direct the performance of the above duty.

3. That the chief of police direct that the police of the city call attention to its performance whenever they shall observe its neglect, and that this action be authorized by a resolution to that effect by the common council.

4. That the collection and destruction of the cocoons of this insect *be confined to those bearing the white egg-masses*, as all others are those of males, or of ichneumonized [parasitized] females, which may be of great service in developing parasites to aid in the work of extermination.

5. That further search be made for the eggs late in the season, when the absence of leaves will permit their detection on limbs and branches where they may not now [in August] be seen.

Most of the cocoons bearing the egg-clusters will be found upon the trees or shrubs upon which the larvæ feed, where they were placed through the instinct of the caterpillar, in order that the young brood, upon hatching, would not have to travel beyond their feeble strength in search of food. The male caterpillar, with no provision of the kind to make, may wander elsewhere in search of shelter, on the edges of overlapping clap-boards of dwellings, beneath the boards and rails of fences, caps of fence-posts, copings of any kind, window-sills, or in any convenient nook. These need not be destroyed, for, upon being opened, they will be found tenantless or, still better, to contain parasitic pupæ.

Jarring.— When the egg-gathering has been neglected and the caterpillars are discovered destroying the foliage of trees of moderate size, as are many of our fruit trees, a sudden jar will bring them down, and if sheets are first spread beneath, they can readily be gathered up and destroyed.

Paris green.— When operating upon larger trees, the caterpillar may be destroyed by showering with Paris green in water. With a force-pump of good construction, the upper branches (usually the first to be attacked, as the larva is noted for its climbing propensity) of many of the horse-chestnuts can be reached from the ground. When the larger elms are infested it will be necessary to ascend the tree by the aid of a ladder and distribute the liquid from the principal branches. The Hydronette, made by Messrs. Rumsey & Co., of Seneca Falls, N. Y., and figured in my first report (page 29, fig. 8), will be found convenient to use in connection with a pail of the mixture hanging upon the left arm. When the mixture is to be employed in this manner, the stirring necessary to prevent the settling of the Paris green would be rather inconvenient, and it would, therefore, be better to substitute London purple, which remains suspended for a longer time in water. A half-ounce of the purple to a pailful of water, or the somewhat stronger mixture of a half-pound to the barrel could be used for the purpose.

Cotton bands.—Bands made of loose cotton batting placed around the trunk, at arm's reach or higher, have become quite popular in some of our cities, as a means of preventing the Orgyia attack. Their indiscriminate use must be condemned. They may be of service, and they may be decidedly objectionable. From what has been written of the natural history and habits of this species, it will be seen that the attack, in almost every instance, proceeds from the *eggs deposited the preceding year upon the tree.* If, therefore, the tree is of so small a size as to admit of its thorough inspection throughout and the certainty that no egg clusters are upon it (which could rarely be attained), then a band applied would prevent the caterpillars which may be wandering in search of food from ascending its trunk and feeding upon the leaves. But if the eggs or the young larvæ are already upon the tree, the band will prove a positive evil. In the event of the caterpillars having so thinned the foliage that a better feeding ground is desired, the cotton barrier encountered in their attempted migration would turn them back to resume and complete their destructive work. With even fewer numbers upon the tree, the barriers would prevent the mature caterpillars from descending to the rougher bark of the lower part of the trunk, in the crevices of which they prefer to build their cocoons, and would confine them to the branches, where it would be difficult to discover the egg-clusters, and to reach them for their destruction as previously recommended.

A New Form of Orgyia Attack.

In the summer of 1883, contemporaneously with the first appearance of the *Orgyia* attack upon the foliage, between the 10th and 15th of June, the sidewalks, streets and public parks in Albany, wherever the white elm (*Ulmus Americanus*) was growing, were observed to be sprinkled with newly fallen leaves. They continued to drop in increasing number until toward the close of the month, when, in many places where they had been permitted to lie undisturbed, they completely covered the walks or ground.

Upon taking some of them up for examination, they were found to be attached to the tips of the twigs and to comprise nearly all of the new

Fig.11.

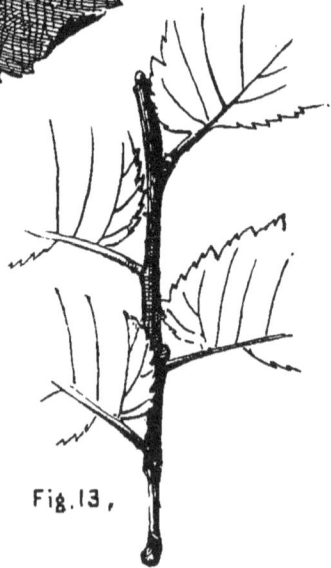

Fig.13,

growth of the season. The pieces were from two to three inches in length, each bearing from four to ten fresh uninjured leaves. It was evident that they were not being broken off by unusually high winds, for even in the absence of winds, each day continued to add to their number and to increase the abundance of the fall.

Making critical observation for the discovery, if possible, of the cause of so unusual a phenomenon, it was noticed that from above the point at which the tip had been broken, the bark was entirely removed for an extent averaging one-tenth of an inch, presenting the appearance shown in Fig. 11. The manner of its removal showed it to have been eaten by an insect. The suggestion was made to me that it was the work of some small insect of similar habits to those of the twig-girdler, *Oncideres cingulatus* (Say), but the closest examination failed to show either scar or egg within the tip.

From the character of the injury, together with the abundant presence of the caterpillar upon the trees at the time, and of no other observed depredator, I believed that it was the work of the *Orgyia*. If so, it was of especial interest, as showing a new habit developed, for this form of attack had never been recorded of the insect. To verify the belief, after ascending some trees and examining branches within reach from windows, I went upon a house-top, where the limbs of a large elm, projecting over the roof, gave an excellent opportunity for examination. The larvæ were abundant upon the tree ; the flat roof was strewn over and heaped in corners with the broken-off tips ; very many girdled tips still held their place on the tree ; and after careful search, Orgyia larvæ were discovered in the act of eating the bark at the girdled points. From later observations, it appeared that the girdling had at this time nearly ceased.

The following explanation of the cause of the falling of the girdled tips seems a rational one. Upon the eating away of the bark by the Orgyia caterpillar, the wood rapidly dried from its exposure to the air and arrest of circulation, and soon became so brittle that from a moderate swaying of the branches, the weight of a half dozen or more of large succulent leaves would occasion the breaking-off of the slender twig — often not exceeding, in its dried state, the diameter of an ordinary pin.

For the occurrence at this time of this novel form of Orgyia attack, I can only offer the following as a plausible explanation : The spring had been remarkably cold, and as a consequence, the development of the foliage had been delayed to quite beyond the ordinary time. The sudden advent of warm weather caused a corresponding sudden start in vegetation, followed by a vigorous growth, and the young tips of the elm would, as the result, be unusually tender. The particular feeding ground of many of the lepidopterous larvæ is known to be selected only after repeated tastings and rejections of such portions of their food-plant as they traverse, and a final acceptance of that most agreeable to them. By a process like this the Orgyia may have made the discovery, that just at the commencement of the new growth, as the result of the seasonal conditions above mentioned, there was concentrated in the tender bark nutriment far more acceptable to it than that offered in the leaves, upon which alone it had hitherto been accustomed to feed. As the bark hardened with the advancing season, it would cease to be desirable for food.

During my absence from the city for most of the months of July and August, the following observations upon the falling tips and the Orgyia were made, at my request, by Mr. W. W. Hill:

July 6. Tips of four to six leaves falling, larvæ — some spinning.

July 10. Many larvæ spinning — a female moth ovipositing.

July 11. A few tips observed.

July 12 and 13. Tips falling fast after a heavy rain — moths ovipositing.

July 14–21. Tips falling — male and female moths observed.

July 22–31. Larvæ, cocoons and moths — many tips falling.

Aug. 1. Larvæ, cocoons and tips observed.

Aug. 2–11. Tips falling, but no larvæ or moths seen.

Aug. 18. Tips still falling.

On my return to Albany for a few days, on the 21st of July, most of the tips then falling and many of those upon the ground presented a new feature. The breaking, instead of being at the base of the girdling, just above the commencement of the new growth, was, in these, at the preceding node, covering the growth of the former year, as represented in Fig. 12. As a rule, the twigs showed a greater diameter at their decorticated portion, compared with those of the earlier fall, and the leaves attached to them had been all more or less eaten by the Orgyia. Their greater strength had permitted them to remain longer upon the tree, and until the death of the preceding internode, which soon followed the arrest of the circulation — its starvation ensuing — it being unprovided with leaves through which a circulation could still be maintained. When dead, a slight movement of the branch by the wind, or even the weight of the terminal leaves would be sufficient to disconnect it at its lower and weaker node. In a few instances, as shown in Fig. 13, where the girdling had been at a little distance above the node marking the commencement of the present year's growth, the separation had been at this point, while others separated in this manner, instead of the

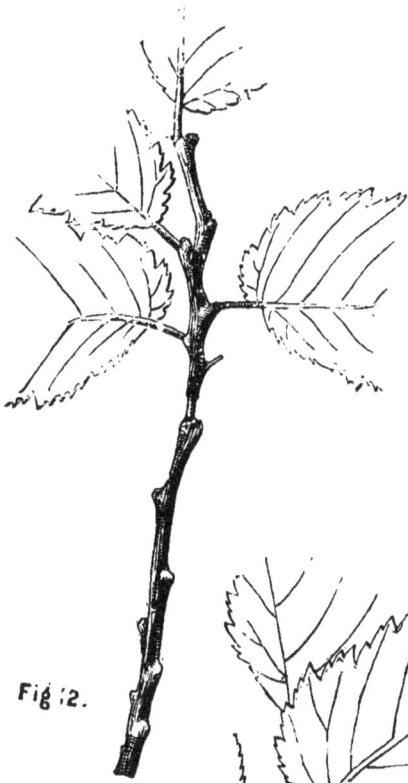

Fig 12.

Fig.14.

Fig.15

narrow girdling band, had had the bark irregularly removed for the extent of an inch or more, as in Fig. 14. All of these later falling twigs showed the interval that had elapsed between the injury and the fall, in that the roughened edges of the bark left by the gnawing had healed over with the peculiar roughened and rounded enlargement following the deposit of the reparative material under such conditions. Some of the twigs gathered gave excellent illustration of the ascent of the sap through the outer wood, and its return, after assimilation in the leaves, through the inner bark. In one instance, where the leaves were unusually large, the descending sap, arrested at the girdled point, had built up structure in the tip until its diameter was more than double that of the starved internode below, while the immediate point of the arrest was quite enlarged from the material there deposited, as shown in Fig. 15.

This peculiar attack did not extend to the other principal food-plants of the Orgyia, as the horse-chestnut, maple, apple and plum, nor would it be expected to occur in connection with growth and structure so different from that of the elm.

The same attack upon the elms, and to about the same extent, was noticed by me in Troy, N. Y., six miles north of Albany. It has not been reported elsewhere, although it probably extended to other localities where the Orgyia abounded under similar climatic conditions. It was not observed in the city of New York, although the caterpillar was very abundant there, and through its excessive ravages, attracted great attention, as has been already referred to.

Plusia brassicæ Riley.

The Cabbage Plusia.

(Ord. LEPIDOPTERA : Fam. NOCTUIDÆ.)

RILEY: 2d Rept. Ins. Mo., 1870, pp. 110–112, f. 81 (orig. descr. and notice); American Entomol., iii, 1880, p. 200 (cannibalistic habits); Ind.-Suppl. Mo. Repts., 1881, pp. 77–8 (descr.); in Papilio, i, 1881, p. 106 (differences from *P. ni*); id., ii, 1882, p. 43 ; in Rept. Commis. Agricul. for 1883, pp. 119–122, pl. 1, figs. 2, 2a, pl. 11, f. 2 (habits, nat. hist., remedies, etc.).

THOMAS : 7th Rept. Ins. Ill., 1878, p. 229 (habits and descr.); 9th Rept. do., 1880, p. 40 ; 10th Rept. do., 1881 (figs. from Riley, brief descr. of larva).

GROTE : in Bull. Buff. Soc. Nat. Sci., i, 1873. p. 147 (as *P. ni*) ; id., ii, 1875, p. 30 (catalogued as *P. ni*) ; in Canad. Entomol., vii, 1875, p. 205 (as *P. brassicæ*); in Papilio, i, 1881, p 127 (may be *P. ni*).

BETHUNE : in Rept. Ent. Soc. Ont. for 1871, p. 5, f. 93, 1872 (from Riley).

PACKARD : in Hayden's 9th Rept. U. S. Geolog.-Geograph. Surv. Terr., 1871, p. 752, figs. *a, b, c* (from Riley).

12

SPEYER : in Stett. Entomolog. Zeit. for 1875, pp. 165, 166 (compared with *P. ni*).
LINTNER : in Colvin's 7th Rept. Surv. Adiron. Reg. N. Y., 1880, p. 399 (as *P. ni*,
    visiting flowers in Ill.); in Count. Gent., xlvi, 1881, p. 711 (general notice);
    1st Rept. Ins. N. Y., 1882, pp. 65, 156 (remedy and parasite).
HOWARD : in Bull. No. 3, Div. of Entomol.— Dept. Agricul., 1883, p. 20 (effects of
    pyrethrum on larvæ).

The increasing destructiveness of this insect within the past few years has brought it prominently into notice, and in response to the many requests made for information in regard to it, its natural history has been ascertained and published, together with the means by which the excessive injuries from it, threatened, may be largely averted.

Examples of the larvæ, pupæ and the moths were received from Dover, N. J., during the latter part of October, with the statement that they had been for some time past, and still continued to be, very destructive to cabbages and Swede turnips, defying the remedies with which the larvæ of the cabbage butterfly, *Pieris rapæ*, had been successfully combatted. The moths had been observed depositing eggs upon the cabbage soon after sunset. It was desired to know if they were an old or a new enemy, and if it was probable that they would continue to be formidable, or were they developed in their present numbers by the recent heat and drouth.

Description.

Larva. — The caterpillar, shown at *a* in Fig. 15, is of a pale-green color, delicately lined in white, with some small white spots, each of which bears a short hair, usually blackish. The head is small, flattened and shining. The body is slender, deeply constricted at the joints, gradually increasing in size from the head to the eleventh segment, where it is rapidly contracted and slopes abruptly to the pair of contiguous anal prolegs : besides these, there are two other pairs of prolegs, and the three pairs of true legs. Its length at maturity is nearly one inch and a quarter.

The method of walking of these caterpillars is rather unusual among the *Noctuidæ*, in that they loop the body after the manner of

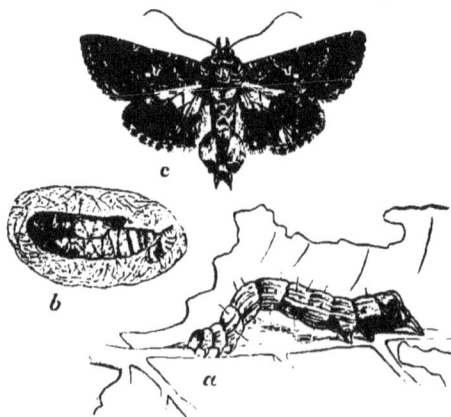

FIG. 15.—The Cabbage Plusia, PLUSIA BRASSICÆ: *a*, the larva; *b*, the pupa within the cocoon ; *c*, the male moth.

the "measuring worms " or "geometers " of the *Geometridæ*, as represented in the figure; but unlike most of the geometers, they are tapering, and not cylindrical throughout. This motion is the consequence of their being unprovided with prolegs on the sixth and seventh segments —having only three pairs of abdominal legs instead of the usual number of five.

Pupa and moth. — The chrysalids or pupæ into which the larvæ transform are contained within loosely-spun silken cocoons, permitting the pupæ to be seen through the threads, as shown at *b*. Several of the larvæ received were spinning their cocoons October 25th, while others were apparently but about half-grown. The pupæ are of a light yellowish or green color, with a projection on the central portion of the lower side, indicating the unusually long proboscis of the moth, for which provision is made in its extended case.

The front wings of the moth are dark grey, almost brown ; the indistinct transverse lines are pale yellowish ; the more conspicuous markings are two small silvery spots near the center of the wing, either united or close together, of which the outer one is oval and the inner U-shaped. The hind wings are smoky outwardly, and yellowish toward the base. The expanse of wings is about one inch and a half.

The male moth shows a peculiar feature in two tufts of fawn-colored hairs, which spring from the sides of the abdomen behind the middle, and meeting on the back. In the figure of the moth, at *c*, this feature has been given unusual prominence, in comparison with the examples seen by me.

For the detailed description of the moth, which might be required for its separation from some closely resembling species, when captured elsewhere than in connection with its larva or general food-plant, reference may be made to its original description in the 2d Missouri Report, pages 111–12.

Food-Plants.

The insect has become a well-known cabbage pest, and its specific name indicates its favorite food-plant, upon which it usually occurs, but it is far from being confined to the cabbage. Its list of food-plants is being continually extended as the result of further observations, and already it embraces kale, turnip, tomato, lettuce, celery, mignonette (*Reseda*), dandelion (*Taraxacum*), dock (*Rumex*), *Crepis*, *Chenopodium*, clover (*Trifolium*), German ivy (*Senecio scandens*), Japan quince (*Cydonia japonica*).* It seems to have developed a particular fondness

---

* Riley : *Rept. Comm. Agricul.* for 1883, p. 119.

for celery , as it is reported as having nearly destroyed a celery patch near Bladensburg, Md.

Distribution and Injuries.

The species occurs in most of the Middle, Southern and Western States, from New York along the Atlantic States, the Gulf States, and the Mississippi valley States to Missouri and Illinois. In the Southern States its injuries to cabbages are quite severe. Its depredations are seldom serious in the Middle States, except in times of drouth and continued hot weather, which seem almost essential to its presence. It has never, to my knowledge, occurred in New York in numbers capable of inflicting much harm, and indeed the moth is regarded as a rarity by our collectors. Within the last five years several of the moths have been captured near Albany, and it is not improbable that the species, as in the case of the harlequin cabbage-bug, *Murgantia histrionica*, may be gradually working its way northward and adding to its permanently occupied territory. It does not yet appear in careful lists of Lepidoptera occurring in the State of Maine. It has been recognized among collections made in California.*

Its Resemblance to an European Species.

It approaches so nearly to an European species, *Plusia ni* Hübn., that its specific difference has been questioned and given occasion to considerable discussion among some of our writers. But as Dr. Speyer, a distinguished Prussian entomologist, whose studies of the allied forms of Lepidoptera in Europe and America have been so critical and just as to have been generally accepted, has carefully compared the two, and believes them to be distinct forms, they are so received, at present, by our leading entomologists.

Number of Broods.

In its more northern extension there are two annual broods, for, from larvæ taken in August, after about two weeks of pupation, Dr. Thomas has had the moths emerge on the 1st of September, which deposited their eggs for a second brood in October. In the Southern States there are probably four broods, for Mr. Grote took examples of the moth in Alabama, during the last of February. Some of the European Plusias are recorded as having three broods a year, although most of them are believed to be single-brooded.

---

*It is reported as appearing, in 1884, in Ramsey and Hennepin counties, in Minnesota, and proving almost as injurious to cabbages as the white cabbage butterfly *Pieris rapæ* (O. W. Oestlund).

Cannibalistic Habits.

An instance has been given by a correspondent of the *American Entomologist*, where this species manifested a cannibalistic propensity. Two of the caterpillars having been inclosed in a box with two of those of the cabbage butterfly, *Pieris rapæ*, and three of another cabbage pest, *Pionea rimosalis*, upon opening the box three days thereafter, it was found that the Plusias had killed and eaten all the others. Should this love for its kind develop into a habit as strong as that existing in the corn-worm, *Heliothis armiger* (see First Report, pages 119–120), it might become serviceable in reducing the number of its associate depredators on the cabbage.

Remedies.

When not extraordinarily abundant, this insect may be kept in check by a moderate amount of labor expended in hand-picking; for as the caterpillars usually rest with their body looped in ∩-shape, they are conspicuous objects, and may be readily discovered and gathered. They may also be destroyed by sprinkling the plants with hot water, of a temperature not exceeding 150° Fahr., if applied through the nozzle of a common sprinkler; but if distributed in the finer spray of the rose of a force-pump, a temperature as high as 170° should not injure the leaves. But probably the best application, the easiest and the most satisfactory, would be that of pyrethrum. A tablespoonful of good fresh powder, diffused through two gallons of water and sprinkled over the plants, would destroy the larvæ — the more quickly if employed while they are still immature.

A preventive measure that would amply repay the effort would be to capture the moths when they are abundant, in a net, as they hover about the plants at dusk in readiness for the deposit of their eggs. That they are so abundant at times as to be easily captured in this manner is evident from the statement made by a gentleman from the District of Columbia, in commending this method, that in a long-continued drouth "*the moths came in swarms*, and afterward come the worms." Catching the moths in their peculiar flight could easily be made a pleasant pastime for the children of a family, while it might serve the additional purpose of leading them to the fascinating study of the habits of the insect world. So simple a thing as an insect net, might thus, under proper direction, become one of the most important of household implements.

Plusia dyaus Grote.

(Ord. LEPIDOPTERA: Fam. NOCTUIDÆ.)

GROTE: in Canadian Entomologist, viii, 1875, pp. 203-4.

Several of the caterpillars of this beautiful moth were received in the month of February, from Dr. R. H. Sabin, of West Troy, N. Y. They had been taken from his conservatory, where they had been proving quite destructive to a number of the plants. They had first been discovered upon a heliotrope which was nearly destroyed before the attack was noticed. They were picked off by hand and killed, as they were of large size — about an inch in length — and readily found, both by the eaten leaves and the masses of excremental matter adhering to the foliage below them. Subsequently others were discovered, feeding quite as greedily upon different species of *Geranium*, brookmansia, easter-plant (*Eupatorium ageretoides*), stevia, etc. They seemed especially fond of the Wandering Jew (*Tradescantia*), the succulent leaves of which were rapidly consumed with evident relish.

The insect had never before been observed by Dr. Sabin, nor, indeed, any such formidable mid-winter attack in his conservatory — the season in which we naturally expect to enjoy immunity from insect ravages.

The Caterpillar.

The caterpillars were innocent-looking creatures, with their small head and attenuated form, in their pale-green garb, delicate white linings and transparent skin, through which the pulsation of the dorsal vessel (the heart) was plainly to be seen. Their peculiar looping movement in walking is like that of *Plusia brassicæ*, as all the species of the genus *Plusia* are provided only with six pairs of legs in lieu of the normal number among the *Noctuidæ*, of eight.

As the caterpillar has not been described, the following detailed description of it is given :

Length at maturity, from 1.3 to 1.5 of an inch ; regularly increasing in size from the head to the penultimate segment. The head is round, flattened in front, and of a shining apple-green color ; the mandibles are black-tipped ; the labium pearly white ; the clypeus transparent, disclosing internal organs; the ocelli are black. A few hairs are observable upon the sides of the head which are nearly as long as its diameter. The breadth of the head is about one-fourth that of the eleventh (the broadest) segment, and nearly equal to that of the first.

The body is clear apple-green in color upon the sides and beneath, and whitish-green over the back. The dorsal vessel is apple-green with its borders well-defined by white lines, contracted at the incisures and considerably enlarged centrally upon each segment. Below the dorsal vessel is a whitish shading, inclosing the anterior of the trapezoidal spots, and extending nearly to a pale whitish-green subdorsal line which is angulated on each segment beneath the posterior trapezoidal spot; there are also two lateral similar colored lines above the spiracles, the lower of which runs nearly straight, while the upper one is quite toothed or crinkled. The spiracles are of a pale orange color. Each of the usual piliferous spots bears a pale-colored hair of the average length of one-fourth the diameter of the body. In addition to these, there are a number of corneous yellowish dots scattered, or grouped in short rows, upon the sides. The body slopes abruptly from the crown of the eleventh (penultimate) segment to the anal pair of prolegs. The legs are long, watery-green, and transparent; the prolegs are dull green with their plantæ tinged with red.

In one of the examples, the slightly elevated piliferous spot above the spiracle on each segment, is black, as are also all of the like spots on the head, of which there are four conspicuous ones arranged in a rhomboid. All of the setiferous spots of the first segment are also black, as are the two dorsal ones on segments two and three; those on segment eleven are annulated with black.

The caterpillar is a rapid and seemingly a greedy eater, and is not easily disturbed when feeding. Its large excremental pellets are distributed over the leaves to which they adhere. At times it assumes the favorite position of many of the *Geometridæ* when at rest. Sustaining itself with its prolegs, it elongates the detached portion of its body and extends it at a considerable angle with the leaf or stem — the three pairs of true legs scarcely visible from being folded upon one another and upon the body and directed forward in range with the head similarly projected.

Transformations.

The larvæ were brought to me on February 10th, when they were nearly full grown. They were fed upon the leaves of the Wandering Jew. Two of the number spun up in slight cocoons made of fine threads within the leaves on the 13th. The cocoons were of so slight a texture as hardly to deserve the name when compared with some of the architectural marvels which many species of caterpillars construct for shelter and protection during their long months of pupation — firm as parchment, double-walled to exclude wet and to regulate temperature, with an accurately adjusted lid in some cases for the escape of the inclosed insect, and with other wonderful contrivances for, and adaptations to, special purposes. It consisted only of a few loose silken threads for support and a thin sheet for covering, thrown across and within the upturned edges of a leaf of their food-plant. Through this thin webbing the caterpillar could be seen for two days after the spinning, when, by gradual contraction and finally throwing off the caterpillar skin, it assumed the pupal form. The pupa could be watched day by day growing darker in color as the insect within was taking form and substance, until twenty days had passed, when the beautiful moths, one after another, emerged.

Beauty of the Plusia Moths.

Guenée, in his *Histoire Naturelles des Insectes*, thus discourses upon the beauty of these moths:

The perfect insects are, without exception, most beautiful. Some of them claim our admiration from their satiny plates in bronze or reddish-brown, of which the most conspicuous precedes the terminal border, and reaches, in gradual contraction, the median space behind the basilar line. But the larger number, in addition to these plates which occur with almost all, bear a particular ornament. This ornament consists in one or two spots placed beneath the cell, peculiar to this genus and not answering to any of the three ordinary spots. These spots are colored with a matter which imitates polished gold or silver in color and brilliancy. They are slightly raised upon their borders as if some drops of these rich colors had fallen upon the wing and become depressed in the middle in drying. But, like all the rest of the wing, they are composed of imbricated scales, of which the exterior ones form a hem in rounded outline, instead of being transversely disposed as are all the other scales of the wing, even those which form the metallic portions often as brilliant as are these particular spots.

But it is not only by their brilliancy that these spots are remarkable; it is also by their form, which is not less essential, and which serves us potently in distinguishing the species. Sometimes they are but simple rounded plates, sometimes two contiguous points; more frequently the posterior one has the form of an oval point, while the anterior is shaped like a hook more or less open, resembling a U or a Y, or yet an interrogation point wanting its dot beneath. Finally, the two spots are bound together, and then the posterior sign takes the form of a tear or a drop, and the entire figure represents a *gamma* or a *lambda* reversed. Both of these and other names of letters and characters have been called in requisition for designating the species of the genus *Plusia*, although these alphabetic signs are not always very striking. (*Noctuélites*, ii p. 325–6.)

The Species Doubtfully Named.

I have referred this moth, for the present, to *Plusia dyaus*, in deference to the determination made of it by Mr. Grote, although probably done without comparison with his type or other examples. The differences observable in the specimens in my possession, when compared with a single example of *P. dyaus* (not in perfect condition, it should be stated) have led me to doubt the correctness of the determination.

In that species, the inner silver spot is subrectangular, tending to form the three sides of a square, while in this it is distinctly rounded where it connects with (or almost touches in one example) the outer tear-shaped character; the two together having much the appearance seen in *Plusia ou* and *Plusia precationis*. The inner transverse line is more nearly straight in this in its course from the inner margin; and there are other differences that need not now be indicated.

Mr. J. B. Smith, whose careful studies have made him our accepted authority upon the *Noctuidæ*, to whom I have submitted an example of the moth with the request that he would pronounce upon it, has returned answer that it is not *P. dyaus* and that he cannot satisfactorily identify it with any species ; but, in consideration of the differences in the determinations of the Plusias in different collections, he would not like to pronounce definitely upon it until after study of the types of all the doubtful forms.

If, upon further study, the present form proves to be new to science, I would propose for it the name of *Plusia culla*.

Amphidasys cognataria Guenée.

The Currant Amphidasys.

(Ord. LEPIDOPTERA : Fam. GEOMETRIDÆ.)

GUENÉE: Hist. Nat., Ins.-Lep., ix, 1857, p. 208, No. 312.
WALKER: List Lep. Heteroc. Brit Mus., 1860, xxi, p. 307.
PACKARD: Guide Stud. Ins., 1869, p. 322 (brief notice); Mon. Geomet. Moths, 1876, p. 413, pl. 11, f. 4 (of moth).
BOWLES: in Canad. Entomol., iii, 1871, p. 11 (descrip. of stages).
SAUNDERS: in Rept. Ent. Soc. Can. for 1871, p. 38; Ins. Inj. Fruits, 1883, p. 349–50, f. 363 (of moth).
LINTNER: in 26th Rept. N. Y. St. Mus. Nat. Hist., p. 166 (larva descr.); in Count. Gent., xlvii, 1882, p. 785 (on apple tree).

13

The popular name given to this insect of the "Currant Amphidasys" is in consideration of its having been found to be more destructive to the currant than to other of the food-plants upon which it occurs.   On one occasion a small black currant bush was observed as presenting a peculiar appearance, as if only the bare stalks of the leaves were remaining upon the branches.   On closer examination the supposed stalks proved to be the bodies of caterpillars of this species, resting in their customary manner, clinging to the branches with their anal legs, and holding their bodies extended (Bowles, *loc cit.*).   Twenty-four of the caterpillars were collected from the one bush.

This species, as appears above, belongs to the family of *Geometridæ*, so called from the peculiar gait that the larvæ have in walking.   The larger number have but ten legs (a few have twelve or fourteen) instead of the ordinary number belonging to caterpillars, viz., sixteen; of these, the three pairs of the front segments of the body are placed closely together and the two hinder pairs quite at the other extremity.   At rest they clasp the twig upon which they are placed, with the two terminal pairs of prolegs, extending the body rigidly outward at an acute angle with the twig, with the front legs folded so closely together and against the head as not to be visible.   In this attitude they are scarcely to be recognized as living objects.   Their obscure colors and some rugosities of surface make them resemble closely a piece of dried twig, for which they are often mistaken.   In walking, the head is brought downward, and the branch is seized by the front legs.   The body is then arched upward in a loop like the Greek letter *Omega*, and the hinder legs brought close to the front ones.   The front legs are again thrust forward as far as possible for another grasp, and thus successive spaces are measured off, which are frequently about one inch in length; and from this the name of "inch worms" has been given to them, in addition to "Geometers" and "loopers."

The Caterpillar.

The insect has not, up to the present, I believe been figured in its larval stage, and we therefore take pleasure in presenting a figure of it drawn by Miss Emily L. Morton, of Newburg, N. Y.   It is represented as feeding upon the honey-locust, *Robinia pseudacacia*, which in that locality seemed to be its favorite food-plant.

From two of the larvæ, obtained by me in September, nearly full-grown, the following description is taken :

Length 1.6 inch; greatest diameter toward terminal segments, 0.2 inch, gradually tapering thence anteriorly.   Head, dull yellowish, subquadrate, but excavated above into two horn-like protuberances (with reddish-brown granulations) which are brown posteriorly.   Body, dull

FIG. 16. — Larva of the Currant Amphidasys, AMPHIDASYS COGNATARIA.

brown with a greenish tinge, somewhat bluish laterally and on the last
segment. On segments three to seven, a subdorsal white spot on the
anterior of each. Collar, along its front edge, with a line of reddish-
brown granulations, ending each side in a brown tubercle. On eighth
segment, two subdorsal pyriform yellowish tubercles, bearing numerous
dark-brown granulations, and on the eleventh segment, two similar more
yellowish ones, more approximate and less elevated and with fewer
granulations; back of these on the terminal segment, the granulations
are whitish. Prolegs upon the ninth and last (twelfth) segment only;
the latter pair spread laterally in walking, are bordered behind with short
whitish spines, and their extreme hinder portion beneath the anal shield,
bears two fleshy horns which come together when moving like a forceps,
almost serving as an additional pair of prolegs. Beneath, on segments
five, six and seven, each, a yellowish wart, of which the middle one is

the largest. The stigmata or breathing-pores are broadly oval, and bordered with red. A magnifier shows a short blackish hair upon each of the ordinary piliferous spots. The legs are reddish.

The Moth.

The moth is quite unlike the usual slender-bodied and broad-winged *Geometridæ*, having a short and stout abdomen, in which, together with its proportionate spread of wings and their general shape, it strongly resembles many of the *Bombycidæ*. Indeed until its structural characters are studied, it seems quite out of place in an arranged collection of the *Geometridæ*.

Fig. 17, taken from Saunders' *Insects Injurious to Fruits*, gives good

representation of the insect. The body and the thorax are gray, the latter with a white collar. The wings are gray, dotted, streaked and lined with black. Two black lines bound the central portion of the wings, the outer one of which is strongly two-toothed on the primaries, and one-toothed on the secondaries. Between these lines, the ground is white sprinkled with black, and traversed centrally by a two-lobed shade on the primaries. The antennæ are broadly pectinated in the male; the abdomen short and quite stout in the female. Expanse of wings, two inches.

FIG. 17.—The Currant Amphidasys, AMPHIDASYS COGNATARIA Guenée.

Life-History.

There are two annual broods of this insect. From larvæ collected in August, I have obtained the imago in the following May. From larvæ taken by Miss Morton on the 28th of June, nearly full-grown, and which entered the ground for pupation on the 10th and 12th of July, the moths were produced from the 12th to the 22d of August. The last of these attracted several males during the night following its appearance, and deposited a large number of eggs on the succeeding night. The larvæ from the eggs were fed upon the honey locust, but being only about one-third grown when the leaves fell, they were not matured. They doubtless died from starvation and exposure in the bag in which they were confined upon a locust, where no suitable shelter could be had, for they were observed traveling about within the bag during all the warmer days of October and November.

Other larvæ taken by Miss Morton during the summer of 1882, buried themselves in the ground, from the 16th to the 20th of September, but failed to produce the perfect insect.

Food-Plants.

In addition to the currant and honey locust and maple above noticed, Dr. Packard records as food-plants of the larvæ, gooseberry, the Missouri currant [*Ribes aureum*], and the red *Spiræa* (*Guide to the Study of Insects*, 1869, p. 322). Miss Morton has also collected it occasionally from horse-chestnut. Mr. William Saunders thinks that it feeds on pine, having captured the moth several times about pine wood, and being also of the impression that he has bred it from the pine (*Rept. Entomolog. Soc. of Ontario*, for 1871, p. 39). In the 23d *Rept. of N. Y. State Cabinet of Natural History*, p. 195, I have recorded taking it upon plum. It has also been sent to me from Wisconsin, as quite injurious to apple-trees, which the caterpillars were represented as killing. That they could have caused the death of the trees upon which they fed, is not at all probable. The belief was doubtless of hasty observation. The gentleman sending them, stated; "Every branch having them on, died. My theory is that they suck the sap from the branch. They did not eat the leaves or branch." After the habit of many of these geometrid larvæ, the caterpillars in the above instance, having satisfied their appetites, had probably left their feeding-ground and traveled to some dead limb where they might elude observation, through their wonderful mimicry of a dead twig. In this position it was perhaps but natural that the inference was drawn that they were the cause of the death of the branch upon which they were found.

Distribution.

This species has broad distribution throughout the northern and western States. It occurs from Maine, throughout the New England States into New York, Pennsylvania, into Ohio (Dayton), Wisconsin (Hoy), and Kansas (Snow). In the Dominion of Canada, it is reported from Quebec and from London.

Remedies.

Many of the caterpillars of this family can be made to drop from the branches upon which they are feeding and hang suspended by their thread, by means of a sudden jar upon the limb, when they can be swept off by a stick and crushed. This species will not often be found in such numbers as to be capable of inflicting serious harm. When it does so, if too numerous for hand-picking, it can be destroyed by spraying the foliage with a kerosene emulsion.

Sitotroga cerealella (Olivier).

The Angoumois Moth.

(Ord. LEPIDOPTERA : Fam. TINEIDÆ.)

—— —— RÉAUMUR: Mem. Hist. Nat. Ins., 1736, ii, pl. 39, figs. 18, 19 ; Mem. Acad. Sci. Paris, 1761.

Alucita cerealella OLIVIER: Encyc. Method.— Hist. Nat. Ins., iv, 1789, p. 121.

Œcophora granella LATREILLE: in Cuvier's Règne Animal, 2d ed., 1829.

Alucita cerealella. HERRICK: in Rept. Commis. Pat. —— Agricul., for 1844, p. 166.

—— ? —— ? JUDAH: in Indiana Farmer and Gardener, for Oct. 4, 1845.

—— ? —— ? RUFFIN: in Farmers' Register, for Nov., 1833 (cited by Harris); in American Agriculturist, for Feb. and March, 1847, vi, pp. 52, 93.

Anacampsis cerealella. OWEN: in Cultivator, N. S., iii, for July and Nov., 1846, pp. 208–212, 344–345 and figures (specific characters, injuries, remedies, etc)

Anacampsis cerealella. GLOVER: in Rept. Comm. Pat.—Agricul., for 1854, pp. 67–69.

Anacampsis cerealella. FITCH: in Cultivator, N. S., iv, 1847, pp. 13, 14.

? Ypsolophus granellus. KIRBY-SPENCE: Introduc. Entomol., 6th ed., 1846, p. 129.

Butalis cerealella. EMMONS: Agricul. of N. Y., 1854, v, pp. 254–5 (descr., habits, remedy).

Butalis cerealella. CHENU: Encyc. d'Hist. Nat. — Noct., 1859, pp. 274–5 (general notice).

Gelechia cerealella. STAINTON: Manual Brit. Moths and Butt., ii, 1859, p. 345 (mention).

Gelechia cerealella. CLEMENS: in Proc. Acad. Nat. Sci., Phila., for May, 1860, p. 158 (description); Id., in Stainton's Tineina N. A., 1872, p. 112.

Butalis cerealella. FITCH: in Trans. N. Y. St. Agricul. Soc., for 1861, xxi, p. 813; 7th Rept. (of 6th-9th Repts.) Ins. N. Y., 1865, pp. 129-133, pl. 1, fig. 2 (extended general account).

Butalis cerealella. HARRIS: Treat. Ins. N. Eng., 1852, pp. 392–402; Treat. Ins. Inj. Veg., 1862, pp. 499–510 (general account); Entomol. Corr., 1869, p. 169 (generic discussion).

Sitotroga cerealella HEINEMANN: Schmett. Deutsch. Schweiz., 1870, Band 2, Heft. 1, p. 287.

Butalis cerealella. BETHUNE: in Rept. Ent. Soc. Ont., for 1871, p. 61 (brief notice).

Gelechia cerealella. FRENCH: in Thomas' 7th Rept. Ins. Ill., 1878, p. 266 (brief mention).

Butalis cerealella. SAUNDERS: in Rept. Ent. Soc. Ont., for 1881, p. 5 (brief mention).

Gelechia cerealella. WEBSTER: in 12th Rept. Ins. Ill., 1883, pp. 144–154 (history, parasites, enemies, etc.).

Gelechia cerealella. RILEY: in Rept. Commis. Agricul. for 1884, pp. 345, 350, pl. 6, figs. 2, 3.

From Mr. E. H. Ladd, of the New York Agricultural Experiment Station, at Geneva, Ontario county, the following communication was received, under the date of October 27, 1884 :

I send you a half car of "forty days early" corn containing the larvæ and pupæ of a moth from which the moths are now escaping in the

Museum. Last night these moths came forth from the specimen sent you. Dr. Sturtevant informs me that it has been present the past two seasons at the station. As the specimen of corn sent you shows, should the insect continue to increase it must do an immense amount of damage to the corn of this section in future years. Will you please give me the name of the moth, and something of its natural history; also in what works are to be found the fullest accounts of it.

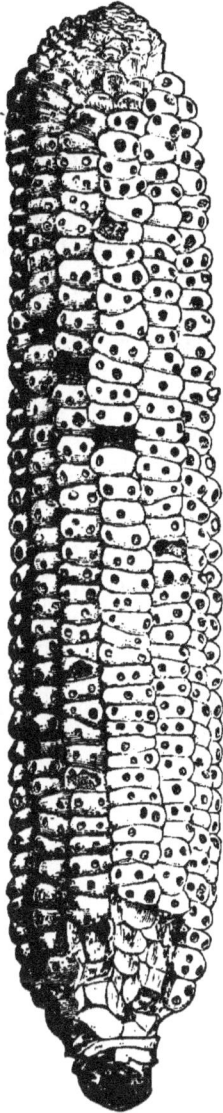

The insect was the Angoumois moth — so named from the canton in France, where its excessive ravages, over a century ago first brought it into general notice. It has long been known as an injurious grain insect in the southern and central portions of the United States, but fortunately it has not proved very destructive in the State of New York or in New England. Originally described by Olivier, in 1789, as *Alucita cerealella*, it has been noticed by several of our entomological writers, in later years, under the name of *Butalis cerealella*, and still later, as *Gelechia cerealella*. This latter genus of the *Tineidæ* having been made the receptacle of a large amount of incongruous material, more careful study is withdrawing from it wrongly referred species, and for the Angoumois moth, Heinemann, in his *Schmetterlinge Deutsch.-Schweiz.*, in 1870, established the genus of *Sitotroga*. This genus appears to be accepted by our authorities in the *Tineidæ*.

Three of the moths were in the box received from Geneva, when opened, and others continued to emerge on following days. The piece of corn, about two inches long, contained eight rows, in which were ninety-six kernels. Of these, sixty-four (sixty-six per cent) contained cells of the insect, as shown by the round smoothly-cut opening of a size less than the head of an ordinary pin, through which the moth had emerged, or by the thin, nearly transparent hull of the kernel covering the cell. Only nine of the cells were open, showing that the occupants had barely commenced to emerge. The pupa-cases were left within the cell where

FIG. 18. — Ear of corn showing the work of the Angoumois moth. SITOTROGA CEREALELLA.

with a magnifier their brown anterior end could be seen.   Upon cutting
into the kernels, the pupæ were found incased in a delicate silken cocoon, held firmly in place by a packing of the large and yellow pellets
of the excrementa which filled the cavity made by the larva.   The
cavity occupies about one-half of the transverse diameter of the kernel
and its entire longitudinal diameter, embracing the germinal portion,
beyond which was a mass of the excrementa adhering to the cob.   In
no instance had a kernel been occupied by more than a single larva,
although the amount of food would have sufficed for two or more.

Dr. Harris, in his account of the depredations of this insect upon
corn, mentions ears in which every kernel had been perforated, with
many of the kernels having three or four holes in each ; but these were
infested ears that had been tied up tightly in paper, and were opened
after the lapse of a year, during which time successive broods had been
developed and were compelled to utilize all of the available food.   An
ear infested to this extent, and presumably from a similar cause, is shown
in Fig. 18, from Prof. Riley's Departmental Report for the year 1884.

It being important to know whether this attack was made upon the
corn in the field, or after placing it in the Museum of the Experiment
Station, examination was made by Mr. Ladd after learning the different
conditions under which it might occur, to determine this point. He subsequently wrote: " From careful examination of the field corn, I have
been unable to find a single instance of its occurrence until brought to
the Museum, and I now think that the whole round of life, in this instance, is within the Museum." This conclusion is confirmed by the
fact that from other ears of corn sent late in October, moths continued
to emerge until into December, which would not have resulted from a
field attack.

History of the Insect.

The ravages of this insect seem to have been noticed in the United
States about the year 1730, when wheat was attacked by it in North
Carolina.   At about the same time, and probably at an earlier period,
it attracted attention in France, for in 1736, the distinguished French
naturalist, Réaumur, gave an account of it illustrated with figures, but
did not give it a scientific name.   Its ravages continued to increase
until in the year 1760, they had become so extensive, that the attention
of the government was drawn to it.   At that time " the insect was found
to swarm in all the wheat-fields and granaries of Angoumois and of the
neighboring provinces, and the afflicted inhabitants were thereby de-
prived not only of their principal staple, wherewith they were wont to
pay their annual rents, taxes, and tithes, but were threatened with famine

and pestilence from the want of wholesome bread."* Two members of
the Academy of Science of Paris were appointed as Commissioners to
visit the province of Angoumois, and investigate the insect. They did
so, and their report was published in the Memoirs of the Academy, and
also as a separate volume for more general distribution, under title of
"Histoire d'un Insecte qui dévore les Grains de l'Angoumois," Paris,
1762, 12 mo.

In 1768 a communication upon it was presented to the American
Philosophical Society of Philadelphia, by Colonel Landon Carter, of
Virginia, entitled "Observations concerning the Fly-Weevil that destroys
Wheat." (Harris.)†

In 1796, the insect was so abundant in North Carolina as to extinguish a lighted candle when a granary was entered at night.

In following years it continued to spread, extending itself into several of the States, particularly those lying within the "wheat belt."
Notices of it were communicated to the agricultural papers and journals, among which are those of Mr. Edward Ruffin, of Virginia, in 1833
and 1847, of Mr. S. Judah, of Indiana, in 1845, and of Mr. Richard
Owen, of Indiana, in 1846, as referred to by Dr. Harris.

Description of the Moth.

The moth, reared from corn, is 0.60 in. (average of seven examples)
in expanse of wings. The head is smooth, the antennæ are nearly as
long as the body, tapering moderately to the tips, each joint being
tipped with black upon the upper side. The palpi are long, curving
backward over the head like horns, the joints quite distinct, the last
one banded with black near its tip. The body and fore-wings are a
dull yellowish or buff color (coffee-and-milk) and of a satiny lustre, especially on the under side of the wings. The front wings are long
and narrow, freckled with black scales, which are thicker toward the tips,
and forming a line along the plait of the wings; the fringe is of a paler
color, long, often trav-ersed with a black line near their attachment to
the wing. The hind wings are blackish, of a leaden lustre, narrow, very sud-

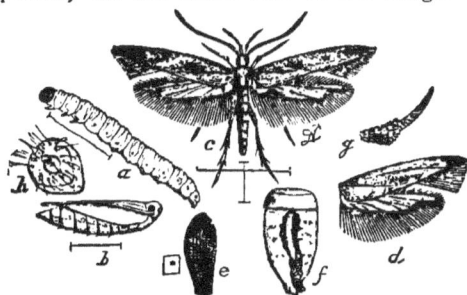

FIG. 19.— SITOTROGA CEREALELLA ; a, the larva ; b, the pupa ; c,
the moth ; d, the wings of a paler variety ; e, the egg ; f, kernel
of corn showing the work of the larva ; g, labial palpus of the male
moth ; h, anal segment of the pupa — all enlarged except f.

*Harris: Treat. Ins. Inj. Veg., 1862, p. 500.
† Mr. E. C. Herrick, in the Report of the Commissioner of Patents for 1844, p. 75, refers
to a communication upon this species from Colonel Carter, in the *Trans. Amer. Philosoph.
Soc.*, vol. i, 1771 ; and another by J. Lorain, in Mease's *Archives of Useful Knowledge*,
vol. ii, 1812.

14

denly contracted to a point near their tip ; the fringe of the same color, surrounding the wings, and along the inner margin broader than the wing itself.   Beneath, both pairs of wings are of a leaden color.   The front pair of legs are blackish ; the hind legs have two prominent pairs of spurs, and are fringed with long hairs.   In Fig. 19, at *c*, an enlarged view of the moth is given.   The excellent figure is from Prof. Riley's Report to the Department of Agriculture, for the year 1884, as is also the preceding one.

At *e*, *a* and *b*, in the figure, the egg, larva and pupa are shown.   For description of these stages the writings cited of Dr. Fitch, Mr. Webster and Prof. Riley may be examined.

Food-Plants.

The most serious injuries of this insect have been to wheat, attacked both in the field and in granaries.   It also occurs in barley, in oats and in corn.   Mr. Glover has seen the moth flying about corn standing in the field, in November, in Georgia, depositing its eggs in the ears.   According to the same writer, it feeds also upon grass-seed (*Entomolog. Index to Agricul. Reports — Anacampsis*, p. 4), although it is difficult to see how it could do so consistently with its concealed habit of feeding, and in all other cases, so far as known, at once burrowing into the grain and feeding and transforming therein.

Life-History.

Under natural conditions there are two annual broods of this insect; but within doors, in stored grain there are more, the number depending upon the temperature of the apartments.

The moths of the first brood make their appearance in May or June, according to the latitude and temperature of the season, and seek the grain upon which to lay their eggs — each moth depositing from sixty to ninety minute eggs, of a bright orange color.   Upon wheat and similar grains these are usually placed in lots of twenty or more, in a line or in an oblong mass, in the longitudinal furrow upon the side of the grain.   The eggs hatching in from four to seven days, the larva at once burrows into it at its most tender portion, and continues to feed upon the interior, packing its excrementa around it in the cavity made by its feeding.   In about three weeks' time it is full-grown, when it measures about one-fifth of an inch in length.   It then eats a hole outwardly for its escape when transformed into a moth, leaving only a thin film, circular in outline, of the surface of the kernel to be pushed off at its final exit.   A thin cocoon of white silk is then spun within the cavity, in which it changes to a pupa.   After a short pupation the pupal case is rent, and the moth emerges, leaving its pupal case within the cocoon.   This is usually in the month of August.

The moths of the first brood soon deposit their eggs for the second brood.   The larvæ proceeding from these continue in the larval stage throughout the winter, to pupate the following spring, and to come abroad as a moth during the months of May or June.

So little has been added to the natural history of this insect since the observations of the preceding century — not to be wondered at, perhaps, in consideration of the better opportunities afforded for its study at that time by its excessive abundance, the distinguished scientists engaged in its investigation, and the government commission under which the studies were conducted — that the account of these observations as briefly given by M. Olivier cannot fail of being read with interest. We, therefore, translate a page (114–115) from the *Encyclopédie Méthodique — Histoire Naturelle — Insectes*, iv, 1789 :

Réaumur has given us the history of another caterpillar which attacks grain, which produces a small *Alucita* that we have named *cerealella*, and which should not be confounded with the one of which we have been speaking [*Alucita granella*].   The caterpillar of *Alucita cerealella* introduces itself even into the substance of the grain, from which it does not emerge except in the state of the perfect insect, to spread itself into the fields, to couple, and to establish a new posterity upon the grains, even before they have hardened.

There appear, in the Memoirs of the Royal Academy of Sciences, for the year 1761, some observations made in Angoumois by Messrs. du Hamel and Tillet, on the caterpillars which caused, in 1760, very considerable damage to the grains of that province.   It seems, from the observations of these distinguished academicians, that the insect often deposits its eggs on the heads of the wheat or the barley before their perfect maturity ; that the eggs are of a beautiful orange-red color; that the larva introduces itself into the grain through a small opening which is found between the beard and the appendages of the sheath ; that the larva grows insensibly without leaving the grain, which serves it at the same time for food and lodging ; that it changes there to a chrysalis, and that it does not come forth but in the state of the perfect insect.

But these caterpillars attack not only the grain in the head, but also in the granaries, as Réaumur, du Hamel and Tillet have observed. When a caterpillar, newly hatched, seeks to pierce a grain of wheat to occupy it, it commences its operations at the lower end of the groove, where the outside is still soft, and consequently more easily penetrated; it spins a slight web which serves to cover it: it pierces the kernel and penetrates by degrees into the interior.   Réaumur has observed that of the grains they attack more particularly wheat, oats and barley, but that they prefer the last, and locate there more readily when they have the choice.   The kernels in which these caterpillars are inclosed appear like the others, since the outside has not been eaten, and that the opening through which the caterpillar entered is imperceptible ; but if different kernels be pressed, the distinction can easily be made of those which have been occupied for some time and those for a short time. The age also of the caterpillar within the kernel, up to a certain point, can be told. If the kernel yield throughout its length under the pressure of the finger, it contains a caterpillar which has nearly attained its growth, or the chrysalis of the caterpillar. If at only a certain part of the kernel it yields to pressure, the caterpillar has not eaten the entire substance, and it has yet to grow. Another means more sure, and shorter, and the more readily to know the kernels attacked by these insects or by the weevil, is to throw in water the wheat or barley : all the eaten kernels will float.

When the caterpillar is hatched, it is so small that a good lens is needed to distinguish it ; it is not more than three lines long when ready for its metamorphosis ; it is smooth and white, its head only is a little brown ; it has sixteen feet, of which the eight intermediary are so small as to be hardly perceptible ; the extremity appears to be bordered with brown hooks, disposed as a crown.

A kernel of wheat or barley contains just the amount of aliment for the food and growth of this caterpillar until its transformation. If one be opened containing a caterpillar ready for its change, only the shell will be found remaining; all the farinaceous substance has been eaten. But before changing into chrysalis, this caterpillar has an important work to do ; it is necessary to provide an outlet which it would not be able to do when it attains the perfect state. The *Alucita*, unprovided with teeth, would never be able to pierce the outside of the kernel for its escape. The caterpillar cuts circularly a piece of the surface, so that it merely holds to the kernel by a portion of its circumference, of which the extent is hardly equal to the diameter of a hair; it does not, however, disturb this piece, so that it does not show so long as the chrysalis is contained in the kernel; it can hardly be told when the insect has emerged from it. After this operation, the caterpillar spins in the interior of the kernel a silken cocoon of a very fine tissue, and changes into the chrysalis. It should be observed that this cocoon does not occupy all the excavation in the kernel — the caterpillar setting off a small space in which it places all of its excrement which it was not able until then to separate.

Messrs. Hamel and Tillet have observed that these Alucitas are usually seen in two seasons, in the spring, as soon as the wheat commences to appear in the head, and these are from the caterpillars that have passed the winter in the grain ; the others appear in the summer, in the neighborhood of the harvest ; these produce the eggs for the first brood, of which we have spoken, and give birth to caterpillars which are to produce the moths of the following year ; some of these may appear during the course of the summer, but the greater part of the number have exactly this order which, however, is sometimes accelerated or retarded by the different temperatures of the air. One thing worthy of remark is, that the moths which emerge in the month of May from grain in the granaries hasten to escape through the windows and to gain the fields ; while those which are disclosed after the harvest show no inclination to escape. It seems that their instinct informs them that they will not find at that time, in the fields, that which would be needed for the well-being of their posterity.

Distribution.

In Europe, its greatest depredations have been in France, but it occurs also in several other portions of southern Europe, extending, at least, northward into Belgium.* Stainton, in his *Manual of British Moths and Butterflies,* records it in England — the larva in grains of barley and wheat, in March and October.

In the United States it probably occurs throughout the entire wheat region. We find special record of it in Massachusetts, Virginia, North and South Carolina, Georgia, Ohio, Indiana and Illinois. It has apparently not extended itself into Canada, for, in 1881, Mr. William Saunders, of London, Ont., stated that it had never, to his knowledge, been found within the limits of that Province.†

Remedies.

When the insect infests stored grains, fumigation in tight vessels with charcoal gas has been recommended. Probably the best agent for its destruction is heat, and we accordingly find the mention of this method by all writers who have considered the means of arresting the ravages of the pest. In France contrivances sometimes called insect mills, have been devised for heating infested grain, and stirring it at the time so that the heat could be uniformly distributed throughout the entire mass. According to Dr. Harris, exposing the grain in a kiln at a temperature of 167° Fahr. for twelve hours has sufficed to kill the insect, and even a very moderate temperature if continued for a length of time — as 104° for two days. Experiments made by Mr. F. M. Webster, of Illinois, to ascertain the amount of heat required, gave these results : " A temperature of 140° Fahr., continued for nine hours, literally cooks the larva or pupa. A temperature of 130° Fahr. for five hours is fatal, as is also 120° for four hours, while 110°, applied for six hours, was only partially effective." It was also found that the highest temperature above mentioned, and even 10° higher (150°), could be borne by wheat for eight hours without impairing its germinating properties.‡ The drying-rooms, arranged with steam pipes, used by many of the large grain dealers of our Western States, would afford every facility for the application of the desired degree of heat to infested grain.

Dr. Herpin, who made special study of this insect during the time of its greatest ravages in France, and who has published largely upon it,

---

* *Catalogue des Lépidoptères de Belgique,* per Ch. Donckier de Donceel, in *Annales de la Société Entomologique de Belgique,* xxvi, 1882, pp. 5-161.

† *Canadian Entomologist,* October, 1881, xiii, p. 198.

‡ *12th Report on the Insects of Illinois,* 1883, p. 152.

proposed exposing the infested material in close tanks to carbonic acid or azotic gas.*

Natural Enemies.

Mr. F. M. Webster has bred from the larvæ of this moth a hymen-opterous parasite, which he has named and described as *Pteromalus gelechiæ* (*12th Report Insects of Illinois*, pp. 151-2). It occurred in considerable numbers — often as many as eight or ten of its pupæ being associated with a single larva. It is probably the same parasite that Mr. R. Owen had discovered in Indiana, in 1846, preying upon these larvæ, of which "some farmers had noticed large numbers among the tailings of the winnowing machine." It was figured by Mr. Owen,† but not de-scribed or named. From the figure, Dr. Harris recognized it as belong-ing to the *Chalcididæ*, and thought that it might be a species of *Pteromalus* (*Insects Injurious to Vegetation*, p. 509).

Mr. Webster also discovered a mite preying upon the larvæ and de-stroying numbers of them. It proved to be *Heteropus ventricosus* New-port (*Linn. Soc. Trans.*, 1850), repre-sented in Fig. 21, after Walker from New-port. The same species has been found by M. J. Lichstentein of Montpelier, France, infesting his breeding cages to such an extent as to destroy nearly all the Hymenoptera, Coleoptera and Lepidop-tera that he attempted to rear. It was described by him as *Physogaster larva-rum*.

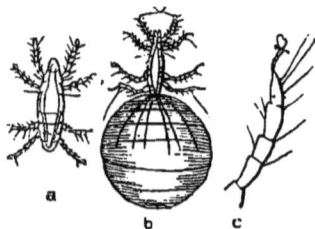

FIG. 21.— HETEROPUS VENTRICOSUS, a mite preying upon the Angoumois moth : *a*, a ma-ture individual ; *b*, female distended with eggs—both enlarged ; *c*, leg greatly enlarged. (After Newport.)

For the interesting observations of Mr. Webster upon the rapid and abnormal development of this mite, and the vigorous attack made by it upon its prey, the reader is referred to Prof. Forbes' *12th Report on the Insects of Illinois*, as previously cited.

DIPTEROUS INSECTS.

Bibio albipennis Say.

The White-winged Bibio.

(Ord. DIPTERA : Fam. BIBIONIDÆ.)

SAY : in Journ. Acad. Nat. Sci. Phila., iii, 1823, p. 77 ; Complete Writings, ii, 1859, p. 69.

---

* *Annales de l' Agriculture Française* for 1838.
† *Cultivator*, N. S. iii, November, 1846, p. 344.

WALSH : in Pract. Entomol., ii, 1867, pp. 45, 83 (habits).
WALSH-RILEY : in Amer. Entomol., i, 1869, p. 227 (habits).
PACKARD : Guide Stud. Ins., 1869, p. 392 (mention).
GLOVER : MS. Notes Journ.— Dipt., 1874, p. 4, pl. 2, f. 7.

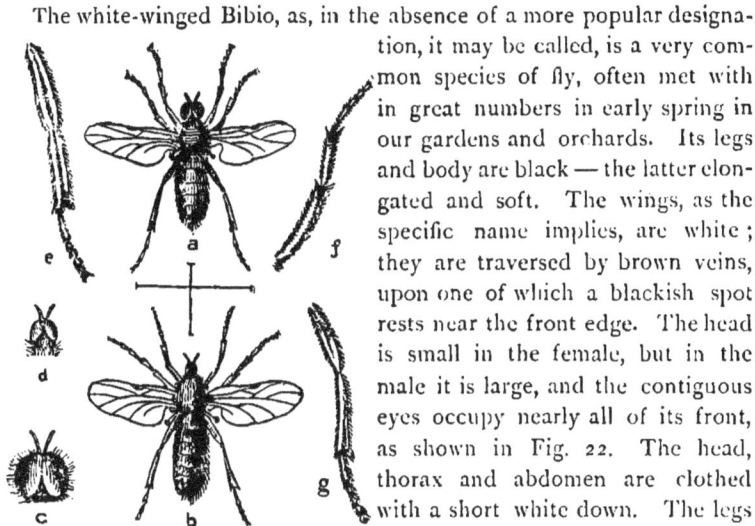

The white-winged Bibio, as, in the absence of a more popular designation, it may be called, is a very common species of fly, often met with in great numbers in early spring in our gardens and orchards. Its legs and body are black — the latter elongated and soft. The wings, as the specific name implies, are white ; they are traversed by brown veins, upon one of which a blackish spot rests near the front edge. The head is small in the female, but in the male it is large, and the contiguous eyes occupy nearly all of its front, as shown in Fig. 22. The head, thorax and abdomen are clothed with a short white down. The legs of the fly, as well as the head, show marked sexual differences, which, however, have been imperfectly represented in the figures.

FIG. 22.—BIBIO ALBIPENNIS : *a*, the male fly ; *b*, the female; *c*, enlargement of the head of the male ; *d*, of the female ; *f*, front leg of the male ; *e*, hind leg of the same ; *g*, hind leg of the female.

The original description by Mr. Say, of the species, and accompanying remarks are as follows ;

B. albipennis. — Black, wings white, with a fuscous stigma. Inhabits Pennsylvania.

Body with cinereous hair ; head above with black hair ; *halteres* [balancers] fuscous ; *scapus* brown ; *nervures* brown ; *tarsi* black-brown, exterior spine of the anterior tibia much larger than the interior ones.

Length, three-tenths of an inch.

This is a very common insect. The wings have a white appearance and are strongly contrasted with the color of the body, and the brown and definite stigma. The posterior tibiæ of the males are much more dilated toward the tip than those of the female.

Family Characteristics.

The family of *Bibionidæ*, to which this species belongs, is of moderate extent, as it contains only about fifty known United States species. Embracing as it does species which are neither injurious to vegetation

nor annoying to man, it is strange that the structural features that char-
acterize it should, in classification, place it between the *Simulidæ* and
the *Culicidæ* — or the " black flies" and the mosquitoes. The insects are
small or of moderate size ; their flight is slow and heavy. They have
long legs, short antennæ, and soft bodies. Their larvæ are cylindrical,
without feet, but furnished with transverse rows of short hairs, which
serve them in progression in the decaying vegetable and excrementi-
tious material in which they often occur. Their pupæ are found within
smooth oval cells in the ground.

The larva of this species is undescribed,* but it probably resembles
that of the " unknown species," to be referred to on a following page.
Its literature is very limited, not having been noticed in the Reports of
Dr. Harris, Fitch, Riley, Packard, Le Baron, Thomas or Forbes. Nearly
all the references to it that we have been able to find in our entomological literature, are cited above and their purport given. A sufficient
reason, perhaps, for its not having received more attention may be that
it is not numbered among our injurious species. Yet the fact that it
has by some writers been regarded as destructive would seem to make
desirable further study of its food-habits, that its economic importance
might be definitely ascertained.

Probably not an Injurious Species.

From the great abundance of these flies which have been observed at
times upon apple and other blossoms, and from the common belief that
nearly all insects are injurious, they have repeatedly been sent to me for
name and habits, in the belief that they were a pernicious species and
that their presence betokened only harm. They are, however, believed
by entomologists to be entirely innocent of inflicting injury to the blossoms upon which they are often found, or to other vegetation. Under
date of May 25, 1877, examples of the fly were sent to me from Utica,
N.Y., where they were occurring in large numbers on potato plants. The
gardener believed that they were making war upon the Colorado beetle,
either by feeding on the larvæ or on the eggs ; but, from what is known
of their habits, there is no probability that they were engaged in so desirable a service — their association with the potato-beetle being simply
from their contemporaneous appearance in the garden, and their usual
sluggish habits inviting repose on any convenient foliage.

Very grave charges, it is true, have been brought against an European
species, *Bibio hortulanus* Meigen. The distinguished naturalist, Ray,
calls it the deadliest enemy of the flowers in spring, and accuses it of

---

*Glover, in *Manuscript Notes from my Journal — Diptera*, 1874, plate vii, fig. 12, gives
the larva and pupa of an undetermined species of the genus, which may be *albipennis*.

despoiling the gardens and fields of every blossom. Réaumur, the noted insect anatomist, saw that, not being provided with mandibles, it was unable to gnaw the buds and petals in the manner ascribed to it, yet thought that it might cause their blight by sucking their juices. Both of these expressed views were doubtless erroneous, and its frequenting buds and blossoms is, upon the best authority, for the innocent purpose of sipping the nectar of the blossoms and the gummy secretions of the opening buds. (*Insect Transformations*, pp. 266–7.)

The larvæ are generally believed to feed almost entirely upon decaying vegetable matter. Some of them which had been found "in bunches" in an asparagus bed, in Missouri having been sent to Mr. Walsh for name and habits, after rearing the fly from them he wrote as follows of the larvæ :

They feed exclusively on dead vegetable substances in a moist and decaying state, and are not very particular as to what that substance may be. Years ago I had a parcel of them feeding on damp leaves in a glass vase, and on putting several dozen of our common "oak apples," into the vase, I was surprised to find that they, most of them, quitted the leaves and burrowed into the oak apples. I have always found them as you did — in large crowds together. They should not be destroyed, as they do no harm in either the larva or the fly state. (*Practical Entomologist*, ii, 1867, p. 83.)

The earlier statements of injuries to growing vegetation were probably from surmise only, and not based on actual observation. Thus, in connection with a plea made for the robin, upon the ground that it destroyed so many of these larvæ — from one to two hundred in a fresh condition having been taken from the stomach of a single bird — it was said : "The larvæ are very destructive, feeding on the roots of plants, and injuring strawberry plats, vines, borders, etc. They live in swarms, perforating the ground like a honey-comb, the fly depositing all its eggs in one place" (Glover, *in Rept. Commis. Agriculture* for 1864, p. 441).

Eaten by the Robin.

The food of the robin, *Merula migratoria*, during the early weeks of its appearance in the spring, consists largely of the larvæ of this fly. Of two robins examined in the month of March by Prof. S. A. Forbes, State Entomologist of Illinois, sixty-seven per cent of the contents of the stomachs consisted of this larva, and the same larva was found by Prof. Jenks, of Brown University, to constitute about nine-tenths of the food of the robins examined by him in Massachusetts in February and March, 1858 (S. A. Forbes, in *Trans. Ill. State Horticul. Soc.*, xiii, 1879, p. 128).

15

Abundance of the Larvæ.

The surprising number in which these larvæ at times occur may excuse the alarm which they sometimes excite.

During the latter part of the winter of 1881–82, they occurred in great abundance in several localities in the town of Morley, St. Lawrence county, N. Y., and occasioned much anxiety, as they were believed by those who first noticed them to be the same insect that had devastated the grass lands of that portion of the State the preceding spring, viz., the Vagabond Crambus, *Crambus vulgivagellus*.

It may not be a simple coincidence that the piece of land where they were the most abundant was the particular field to which in my visit to this town the year before to examine the operations of the Crambus, I had first been taken, that its greatest devastation might be shown to me. Could the unusual amount of dead and decaying grass roots left in the ground from the death, in large part, of the grass the previous year, have induced the deposit of an extraordinary number of the Bibio eggs — the instinct of the parent flies recognizing the presence of the needed food-supply for its larvæ?

After "a very long continued heavy rain on March 1, lasting about fifteen hours, and a steady pour all the time," the larvæ were found lying upon the ground under leaves, sticks and other covering, often in such numbers as to form a mass of an inch in thickness.* They had been observed, but in less abundance, for some days previous to this rain, when, according to a representation made, "several quarts of them could easily have been gathered."

Transformation to the Fly.

Over a hundred of the larvæ sent to me by Secretary Harison of the New York State Agricultural Society, on March 2, were placed upon a dead grass sod in my office, on March 6, when they at once commenced burrowing into it, and soon disappeared from sight. They were doubtless near their pupation, for three weeks thereafter (March 29), the first *Bibio albipennis* emerged — a male. On the 30th, eight males and three females were disclosed; on the 31st, four males; showing that in this species, as in a very large proportion of other insects observed, the male is the earliest to make its appearance.

Is the Species Double-Brooded?

Dr. Packard states of the species that there are two broods a year,

---

*See notice of large numbers of the larvæ of *Allorhina nitida*, one of the "white grubs," appearing above ground after heavy rains, in the *First Rept. on Inj. Ins. of N. Y.*, pp. 238–9.

which appear " in swarms in June and October " (*Guide to the Study of Insects*, 1869, p. 392). It has never come under my observation in the autumn, and I know of no other mention than the above, of its appearance at that time. To an inquiry made of Mr. E. L. Keen, of Philadelphia, who has been paying special attention to the Diptera, of his knowledge of a second brood, the following reply was received :

" I have been giving special attention to the *Bibionidæ*, but I never saw a specimen of *Bibio albipennis* in the fall — in fact not after July, although in May and June there are swarms of them. This last October I took a few specimens of a small black species of Bibio in a sheltered ravine."

An Unknown Species Occurring in Rose-pots.

The larvæ of a species of *Bibionidæ* were found in large numbers in rose-pots in New York city. When received, on the 13th of February, they were already nearly full-grown. They had probably been taken from a conservatory, the artificial temperature of which had hastened their growth.

They measured 0.35 of an inch in length, by about 0.05 of an inch broad. Their shape was nearly cylindrical, without feet; a rounded, corneous brown head of rather more than one-half the diameter of the body ; on each of the segments, a transverse row of short papillæ, and a row of larger ones curving at the tips, on each side of the body.

Their general appearance was so unlike the ordinary dipterous forms,

and so similar in the character of the head and stigmata to some of the Coleoptera that they were at first believed to be of that order.

Within a fortnight after their reception they had all entered the ground to the depth of about three inches, where they transformed

FIG. 23.—Imago and pupa of ASPISTES. sp. ? Twice the natural size.

to pupæ having the appearance shown in Fig. 23. The flies emerged after a pupation of about two weeks. They have a general resemblance to *Bibio albipennis*, except in darker wings, and they are only about one-half its size.

Upon submitting them to a friend who was engaged in the study of Diptera, he referred them to the genus *Aspistes*. There seems to me, however, to be reasons why they may not belong to that genus.

[Baron Osten Sacken, having seen these examples since the above was put in type, has expressed his opinion that they may be but dwarfed *Bibio albipennis*.]

Microdon globosus (Fabr.).

(Ord. DIPTERA: Fam. SYRPHIDÆ.)

Mulio globosus FABR. Syst. Antl., 1805, p. 185. No. 7.
Scutelligera ammerlandia SPIX.: Abhdl. Acad. Muench., 1824, ix, p. 1.*
Parmula cocciformis v. HEYDEN: in Isis, 1823, p. 1247; in H. Schæf. Correspond-
     enz blatt, ii, 1861, 105.*
Microdon globosus WIEDEMANN: Aussereurop. Zweif. Ins., ii, 1830, p. 86, No. 11.
Dimeraspis podagra NEWMAN: in Ent. Mag., v, 1838, p. 373.
Aphritis globosus MACQUART: Dipt. Exot., ii, pt. ii. 1841. p. 13, pl. 1, f. 4.
Microdon globosus. PACKARD: Guide Stud. Ins., 1869. p. 398. f. 17.— GLOVER: MS.
     Notes Journ.-Dipt., 1874. p. 32, pl. 8, f. 20.— OSTEN SACKEN: in Bull.
     Buff. Soc. Nat. Sci., iii, 1877. p. 41; Cat. Dipt. N. Amer., 1878, p. 119.—
     MANN: in Psyche, iii, 1882. p. 379.

This species will serve to illustrate the great variety of forms that prevails in the early stages of the different genera of this extensive family. In Fig. 24 the curiously spherical larva, the puparium and the perfect insect are shown, as given in Packard's Guide. So unusual and remarkable are the larval and pupal forms of this insect, that the latter has twice been referred to a different class of the animal kingdom, and described among the Molluscs, as a land snail, viz., by Hayden in 1823, as a species of *Parmula*, and by Spix in 1824, as a *Scutelligera* (see synonymy above). One of the puparia was received by me in the month of April, from Mr. C. M. Weed, of Lansing, Mich. From having been apparently dried before placing it in alcohol, it was hardly recognizable when it came to hand, but it was identified by Dr. Hagen, to whom I am also indebted for portions of its synonymy. It had been found by Mr. Reed in an old and decaying log, with some ants. Dr. Packard states that it occurs under sticks, in company with shells.

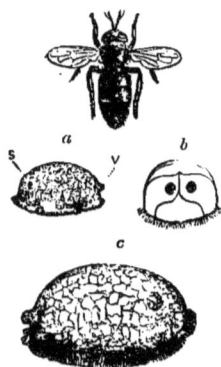

FIG. 24.—MICRODON GLOBOSUS: a, the puparium, s, the spiracular tubercles and r, the vent: b, anterior view of the puparium; c, the larva just before pupation (enlarged from Packard).

Associated with Ants.

From its association in old wood with ants, in the above instance, it is of interest to recall the occurrence of an European species, *Microdon apiformis*, in ants' nests, and also that some of the larvæ of the *Syrphidæ* are known to live parasitically in nests of bees.

Microdon tristis Loew, a New York species, and also extending far northward into British America, has been observed by Dr. Williston while flying about ants' nests.

* Dr. Hagen, in making corrections to the proof of synonymy, has written: "v. Heyden and Spix belong to the European species, *Microdon mutabilis* and not to *M. globosus*, but they are the first authors for the earlier stages."

Schiner states (*Fauna Austriaca*, p. 250), that he has found the larvæ of European species, resembling small slugs, in colonies of *Formica rufa*, and also under logs in cattle pastures.

Hibernation of the Fly.

Mr. B. P. Mann states of the fly, *loc. cit.*: " I remember that, during one or two years, at a certain season, which, as far as my recollection serves me, was in April, I noticed numerous specimens of *Microdon globosus*, issue from a nail-hole in the plastered wall of an apartment in a dwelling-house, as though the flies had passed the winter within the walls of the house." In this manner of hibernation within dwelling-houses, it conforms to a similar habit recently published of *Pollenia rudis* Fabr., one of the *Muscidæ*, which often occurs " in large numbers in the State of New York, in unused apartments of houses, under table-cloths, in pillow-cases and wherever similar snug places of concealment could be found." And it is not uncommon to meet with large companies of these "cluster-flies" in the angles of the ceiling and the walls of unoccupied rooms.

Distribution.

Baron Osten Sacken gives the range of this species as the Atlantic States. It was first described by Fabricius, from examples received from Florida. Dr. Williston has received it from Massachusetts, Connecticut, New York and Pennsylvania and Virginia. Michigan is now to be added to its known localities.

The genus appears to pertain to the eastern part of the North American continent, for while eight or nine eastern species have been catalogued for some time past, it has only been recently discovered upon the Pacific coast in a single species which is as yet undescribed.

Trypeta pomonella Walsh.

The Apple Maggot.

Ord. DIPTERA : Fam. TRYPETIDÆ.)

WALSH : in Amer. Journ. Horticul. for Decem., 1867, pp. 338-343; 1st Ann. Rept. Ins. Ill., 1868, pp. 29-33, figs. 2, 2a.

WALSH-RILEY : in Amer. Entomol., i, 1868. p. 59 (prob. ident. of insect from N. Y.).

LOEW : Mon. Dipt. N. A., Pt. iii, 1873, pp. 265-268 (descr. and remarks).

GLOVER : MS. Notes Journ.— Diptera, 1874, p. 58, pl. 9, f. 14 (br. ref. and authority).

COMSTOCK : in Rept. Comm. Agricul. for 1881–1882, pp. 195–198, pl. 14 (operations, larva, pupa, imago, remedies).

SAUNDERS : Fruit Insects, 1883, pp. 135, 136, f. 143.

COOK : in Count. Gent., xlix, 1884, p. 857 (habits and occur. in Mich.); 14th Ann. Rept. Mich. St. Horticul. Soc., for 1884, 1885, pp. 200–203, figs. 1–3 (general notice).

LINTNER : in Bull. N. Y. Agricul. Exper. Station, lxxv, Dec. 29, 1883 (description, habits, etc.).

In November, 1883, the following letter of inquiry was submitted to me from the Horticulturist of the New York Agricultural Experiment Station, with the request that I would make reply to it :

We are troubled here with a worm that completely destroys our apples. It seems to be different from our common apple-tree worm — not so large, but looks like the apple-tree borer when small. Last year was the first that I was troubled with them. The apples look well on the outside, but when cut they are found to be completely honey-combed by the worm, making them worthless. Last year they worked in my Spitzenberg apples. This year the Spitzenbergs are good, but the worms worked in the Northern Spy, Talman sweeting and Fameuse, and entirely spoiled them. They also worked some in other kinds. The Rhode Island greenings are free from them yet, and the Baldwins are but slightly injured. There is a general complaint this year.

The question which I wish to ask is, are these worms going to spread until it is impossible to raise apples fit to use, or is there some way to stop them ? Last year I took pains to pick up all the poor apples in hopes thus to destroy them, but failed, as they have increased ten-fold from last year. But I do not understand why the Spitzenbergs were ruined last year and are so fair this year. If you can give me any light on the subject, it will be thankfully received.

M. P. JUNE.

BRANDON, VT.

My reply to this inquiry was published in the *Husbandman* [of Elmira, N. Y.] for December, 1883, and also issued, in slips, as Bulletin No. LXXV, of the New York Agricultural Experiment Station. It has been somewhat extended and re-arranged in the present notice.

It is quite probable (the infested apples were not submitted for examination) that the insect whose operations are above stated is the one popularly known as " the apple maggot," and scientifically as *Trypeta pomonella* Walsh. The species was first described and named by Mr. B. D. Walsh, in 1867, in the American Journal of Horticulture.

The Larva.

Operations. — Mr. Walsh writes of its work as follows : " It tunnels exclusively the flesh or pulp of the apple (unlike the apple-worm, which burrows chiefly in the core and the portions immediately surrounding) making therein little rough, roundish, irregular and discolored excavations about the size of peas, which, when several of the larvæ are at work on the same fruit, often run together, so as to render the whole a mere mass of useless and disgusting corruption."

The following is Mr. Walsh's description of the larva, which is given entire, as the report in which it is contained has become quite rare, and it has never, to my knowledge, been republished. With the description and the larva in hand there should be no difficulty in determining the species whenever the burrowing of the apple, as above described, renders its presence probable :

It is of a greenish-white color, 0.15, 0.20 inch long, and about four and one-half times as long as wide, cylindrical behind, with the tail-end squarely docked, tapering on front from the middle of the body to the head ; head pointed, but narrowly excavated (emarginate) in front ; its inferior surface with two slender, bluntish, coal-black hooks projecting in front, where the mouth is protruded [mouth-parts, used in feeding, shown in Fig. 25], at the base of which there is a smaller pair connected with the base of the others like the antlers of a buck's horn ; at the base of the first segment, behind the head, a dorso-lateral, transverse, pale-brown, flattish, rough tubercle [one of the spiracles] ; last segment below with two pale-brown, horny, rough tubercles, each composed of three minute thorns longitudinally arranged ; and above, with two whitish retractile ones, each pair of tubercles transversely arranged.

Fig. 25.— Head of the larva of TRYPETA POMONELLA, in front and side view, showing the mouth-parts and the first spiracle. (After Comstock.)

Prof. Comstock, who has made special study of this insect, has given an enlarged figure of the larva (loc. cit.). which we copy, somewhat reduced, in Fig. 26 at a, and the following descriptive features : The larva averages about one-fourth of an inch in length (0.19–0.27 in.), is footless, white with sometimes a yellowish or greenish tinge. The anterior third of the body tapers slightly to the head — the latter smaller than any of the segments. The posterior two-thirds of the body is cylindrical, having the end obliquely truncate, bearing upon its slope four pairs of tubercles — one pair more prominent than the other.

The Fly.

Trypeta pomonella, in its perfect state, is a pretty fly, shaped not unlike the common house flies that we find upon our windows. The figure given does not present the usual shape of the abdomen in the examples in my collection. In these, it is the broadest quite near the base where the first and second of the four conspicuous white bands should appear, and thence tapers conically to the pointed tip. It may be easily recognized by the peculiar pattern of its wings, by the white spot behind the thorax (the scutellum), and the white bands of the abdomen.   With its wings expanded, it measures across them about one-fifth of an inch. They are "whitish, glassy, banded with dusky, somewhat in the form of the letters I F — the I placed next the base of the wings, and its lower end uniting rather indistinctly with the lower end of the F; the base and the extreme tip of the wing being always glassy " (Walsh).

Fig. 26, *c*, after Comstock, represents the male fly.* The following more particular description is abridged from Dr. Loew's *Review of the North American Trypetina :* The color is brownish-black. The head is pale yellowish with a narrow dark yellow front, and yellow antennæ. The thorax shows four rather narrow longitudinal stripes of whitish pollen, arranged in pairs and confluent anteriorly. The scutellum is white, with black upon the sides and base. The first four segments of the abdomen are broadly banded behind with white pollen — the last segment is without the band. The ovipositor of the female is black, broad, trun-cate, about once and a half the length of the last segment. The legs are mainly clay-yellow, thus : posterior femora black with a clay-yellow tip; front femora clay-yellow with a large, broad brownish-black stripe upon the hind side ; tibiæ and tarsi clay-yellowish. The wings are hyaline, with four black crossbands, the first of which lies near the base ; the last three are connected near the anterior margin and divergent toward the posterior one. Length of body, 0.17 in.; with the ovipositor, 0.19 in.; length of wing, 0.17 in.

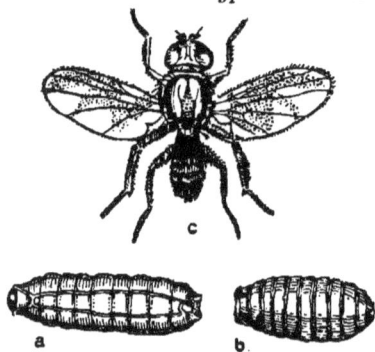

FIG. 26.—TRYPETA POMONELLA: *a.* the larva; *b.* the puparium; *c,* the fly. (After Comstock)

Life-History.

The life-history of this species is briefly this : The fly may be seen about apple trees in July. During the latter part of the month, or in August, it deposits several of its eggs upon an apple near the calyx end, where the fruit may have been already burrowed by the apple-worm of the codling-moth, *Carpocapsa pomonella.* According to Mr. Walsh, the eggs are inserted by the ovipositor of the fly within the flesh of the ap-

*The figure fails to represent correctly the one from which it was taken. From the reduced photograph furnished the printer for a wood-cut, a pen and ink sketch was made and photographed, in which proportions have been changed, veins and other features omitted, and additions made. Figs. 21 and 25 and some others need the same explanation.

ple ; but this must be an error, unless they should be deposited in worm-holes previously made, for the soft, blunt ovipositor could not pierce the peel for the purpose. The presence of the larvæ, of which, sometimes, as many as a dozen occupy a single apple, is seldom noticed until in September. During the autumn they become full-grown, when they leave the fruit through small circular holes cut in the peel, and enter the ground, where, within the contracted larval skin, they assume the pupal state. In this condition they remain during the winter and until the following July — a pupation extending over about nine months.

If some recent observations upon this insect in some of the Western States are reliable, then, in some instances, the larval stage of the insect extends far into the winter. Dr. F. W. Goding, of Ancona, Mich., states that he has seen the larvæ within apples shipped from Michigan, eating the fruit in the month of January, but that they soon after entered the earth and changed to pupæ. Some that had been kept in a cooler room did not change until in March. From the earliest pupæ, flies were obtained about February 1st (*Fruit-Growers' Journal*, Cobden, Ill., April 30, 1885).

Distribution.

The fly is a native species — one of the few insect pests that has not been introduced from abroad. For a number of years before it was detected infesting the apple orchards of the Eastern States and New York, it had been observed in Illinois, feeding upon the native haws or thorn-apples (*Cratægus*) and upon crab-apples (*Pyrus* species). Until very recently no attack from it upon cultivated apples had been noticed in the Western States, but during the present year (1884) Prof. Cook has discovered it in a barrel of autumn apples procured from Shiawasse county, Mich., which were found to be entirely ruined by it. He had also learned that it was quite common in the fruit in and around Lansing, Mich. (*Count.-Gent.* for Oct. 16, 1884).[*]

The most frequent notices of the presence of the insect, and most serious accounts of its injuries, have been received from Vermont. In New Hampshire, in a few localities, it has ruined entire orchards (*Rept. Comm. Agricul.* for 1881, p. 190). In the vicinity of East Falmouth, Massachusetts, it has been reported as very injurious (*1st Rept. Ins. Ill.*, 1868, p. 31). Mr. L. L. Whitman has written me from his resi-

---

[*] Later, Prof. Cook reports its presence in at least six counties in Michigan, and had learned that in 1883, a year before it was noticed in Michigan, it was doing much harm in Wisconsin to cultivated fruit. These attacks were the result of new importations from the East, or the native insect had of late taken upon itself a new and refined regimen. (*14th Rept. Mich. St. Horticultural Society*, for 1884, p. 200-1.)

dence at North Ashburnham, Mass., " I had hundreds of bushels of the
finest fruit rendered worthless by the apple-maggot last year" [1883].
Mr. Walsh received larvæ and pupæ from Connecticut, from which he
reared the fly (*Ib.*, p. 31). In New York State it has proved a great
pest at North Hempstead, Long Island. Mr. Trimble, of New Jersey,
in 1867, stated that it was very plentiful throughout the Hudson River
country, but had not been observed in his State. It has been prevalent
at the Oneida Community, where it was mostly confined to certain varieties of autumn apples (*Rept. Comm. Agricul.* for 1881, p. 196).
Prof. Comstock, of Ithaca, has observed it in one of the orchards of
Cornell University, in only a few varieties, not specified (*Id., ib.*). From
Franklin, Delaware county, larvæ have been reported as "living in the
pulp of the apple, making long winding roads through it and appearing
to come out through the skin" (*Amer. Entomol.*, i, 1868, p. 59). In
Schenectady the fly has occasionally been captured by me upon fruit-
trees, from the 3d to the 27th of July — at different times upon the
leaves of a cherry-tree, where it had probably been drawn for the purpose of feeding upon the " honey-dew " of *Myzus cerasi*. which it in-
cites the aphis to secrete after the manner related by Dr. Fitch in his
First Report on the Insects of New York, p. 65, of *Tephritis melligi-
nis.**

Preference for Early Apples.

As the insect, during its past history, has shown a decided preference
for summer and autumn apples, its attack upon Spitzenbergs, mentioned
by Mr. June, is of interest, as an extension of its sphere of operations.
In Massachusetts the crop of some summer sweets had, for a series of
years, been completely destroyed by it. In New Hampshire, in the
towns of Hancock and Dublin, where it was reported as the most destructive of all apple insects, it was "confined to early apples as soon as
they ripen." From Long Island it is stated that " only in the ripest
apples and in sweet and mellow subacid fruit are they found by us "
(*Rept. Comm. Agric.* for 1881, p. 196).

It seems, however, by no means to be confined to the early fruit, for in
Wallingford, Conn., the Baldwin and some other varieties of winter
apples were infested. A gentleman writing from this place states : " Two
weeks ago [the letter was written shortly after November 12th] we over-
hauled two hundred and fifty bushels of apples that we had gathered and
placed in store for winter use, and of that number we threw out fifty
bushels, most of which had been rendered worthless, *except for cider* [!!!]
or hogs, by the apple-worm or apple-maggot. The apple-worm [*Carpo-*

---

* Is *Ricillia viridulans* R. Desv. (O.S. *Cat. Diptera N. A.*, 1878, p. 182.)

capsa pomonella] by this time has ceased its work, but the depredations of the apple-maggot continue up to the present time, converting the pulp of the apple into a mere honey-comb and rendering another overhauling, soon, indispensable."

As a possible explanation of the statement of Mr. June that Spitzenberg apples were free from attack in 1883, when they had been destroyed the preceding year, the following is offered: The period of time during which apples are liable to attack in the deposit of the eggs may be comparatively brief. At the time when the fly was abroad, in 1883, and engaged in oviposition, the Spitzenbergs, from seasonable peculiarity, may either have been not far enough, or too far, advanced for the reception of the eggs.

Remedial Measures.

In the event of its being found by examination that the infested fruit, to any large amount, falls to the ground with the contained larvæ, it is obvious that much may be done by giving sheep the range of the orchard to feed upon the apples, or, if this be not practicable, then to pick them up and feed them to stock. But, from the statement of Mr. June, it seems probable that the fruit which fell from the trees did not contain larvæ. This would seem to be supported by the fact that I can recall, a few years ago, seeing a variety of early apple (the kind not noted) offered for sale in the Albany market, which, while entirely fair externally, and showing none of the bruises of windfalls, upon being cut into, the central portion was found burrowed after the manner described and semi-pulpy, and still retaining the larvæ — with little doubt, those of this *Trypeta*.

Additional observations are needed upon the time and manner of the larvæ leaving the fruit, and, in the earlier kinds of apples, entering the ground, while in the later and stored varieties it is desirable to learn where the insect goes for pupation. Until these facts are ascertained, with others that are needed to complete the life-history, we will not be able to do much toward mitigating the evil. The attack of the apple-worm producing the codling-moth may be to a great extent prevented by showering the trees soon after the setting of the fruit with Paris green or London purple in water, but the month of July, when the Trypeta deposits her eggs, would be too late for the application of these poisonous substances to the early fruit. At the present our efforts may, perhaps, be directed with the best success toward the destruction of the insect in its pupal state. If it shall be found that the pupation, as a rule, takes place in the orchards, beneath the infested trees, in the ground at the depth of an inch or two, then it may easily be reached there. But, if the pupation follows the storing of the fruit, then the discovery of the retreat of the larvæ for pupation should give us the means for their destruction.

Desiderata in its Life-History.

The injuries from this insect have already been serious in some localities, and their extension is threatened over the Western States where the
species exists but has not yet learned to prey to any very serious extent
upon cultivated apples.   As it may be regarded at present as a local pest,
it should be carefully studied by those who have the opportunity of
observing it.   As a guide to such study, attention is called to the following points, which specially need to be determined :

Are the eggs of the fly distributed over the apple or placed only near
the calyx end?

Do the larvæ occur in apples which have not been perforated by the
apple-worm of the codling-moth or some other insect?

How long a time is required for the larvæ to attain their growth?

How do the larvæ leave the fruit — by several holes through the skin,
through a single hole, or only when the apple has become broken down
from decay?

When entering the ground for pupation, to what depth do they bury?
This could be ascertained by providing them with a box containing a
few inches of earth for their burial.

Are both the early and late fruits similarly attacked by this insect?
It is possible that the larvæ reported in winter apples may be of a different species.

During what months and portions of months are the larvæ to be found
in the apples?

Are the puparia (see *b*, in fig. 26), to be found at the bottom of
apple barrels, or bins in cellars, or between the staves or boards?
Should any doubt exist of the identity of the puparia found under such
conditions, the fly should be reared from them to determine the point.

Other Species of Similar Habits.

The qualified naming, at the commencement of the notice of the
insect of which inquiry was made, was necessary, in the absence of
examples for examination, there being two other species of flies which
attack and injure apples in much the same manner as the apple-maggot.

The first of these is believed to be a species of *Drosophila*. Its operations in early apples were described in Vol. II of the *American Naturalist* (p. 641).   The fly and the
puparium are represented in Fig.
27.   The larvæ enter the apple
either through the calyx end, the
hole bored by the Carpocapsa apple-worm, or the cut of the Curculio. They mature in August and
September, and are followed by

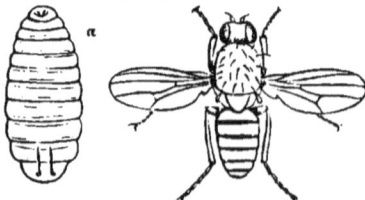

Fig. 27.— An apple-fly. DROSOPHILA, of an undetermined species : *a*, the larva.

another generation. The puparia are to be found in the bottom of apple barrels in cellars, and the flies appear in the spring. The species does not appear to be known to entomologists, and it has therefore been given in the recently published volume of Mr. Saunders on *Insects Injurious to Fruits* as *Drosophila* — - ? In Packard's *Guide to the Study of Insects*, it is referred to (p. 414) as the Apple-Fly. It was also noticed in my *First Report on the Insects of New York* (p. 219), together with the larvæ believed to produce the fly.

The other species is the *Molobrus mali* as named and described by Dr. Fitch in his *Second Report on the Insects of New York* (Reports I and II, 1856, pp. 252-254), but now known as *Sciari mali*. It belongs to the *Mycetophilidæ* which has place next, and is closely allied to, the *Cecidomyidæ* containing the wheat and clover midges and the Hessian Fly. This species has been given the name of the apple midge. Its larvæ and pupæ need not be mistaken for either of the two preceding species, the larva being long, slender, shining, glossy-white, without feet or tubercles (family features, the specific ones, it is believed, have not been recorded); while the pupa is not hidden within an outer case, but shows distinctly the legs, wings and other external features of the future fly, and, moreover, remains within the fruit — the fly emerging from it and escaping through the opening previously made by the codling-moth caterpillar.

INJURIOUS COLEOPTEROUS INSECTS.

Amphicerus bicaudatus (Say).

The Apple-twig Borer.

(Ord. COLEOPTERA: Fam. PTINIDÆ.)

Apate bicaudatus SAY: in Journ. Acad. Nat. Sci. Phila., iii, pt. ii, 1824, p. 319.
Apate bicaudatus. HARRIS: Rept. to Amer. Pomolog. Soc., 1854, p. 7 (brief notice of distrib., etc.).
Bostrichus bicaudatus. FITCH: in Trans. N. Y. St. Agricul. Soc., for 1856, xvi, p. 330; 3d Report. Ins. N. Y., 1859, p. 12, No. 12 (brief notice).
Bostrichus bicaudatus. UHLER: in Rept. Commis. Pat. for 1860, p. 321, 1861 (descr. and remedies).
Bostrichus bicaudatus. WALSH: in Practical Entomol., i, 1865, p. 27, f. 3 (habits and distribution).
Bostrichus bicaudatus. WALSH-RILEY: in Amer. Entomol., i, 1868, p. 80, f. 69 ; p. 206, f. 141.

Amphicerus bicaudatus. SHIMER: in Trans. Amer. Ent. Soc., ii, 1869, pp. viii, ix (in grape vines).

Amphicerus (Bostrichus) bicaudatus. GLOVER: in Rept. Commis. Agr. for 1872, p 118, fig. 8.

Bostrichus (Amphicerus) bicaudatus. RILEY: 4th Rept. Ins. Mo., 1872, pp. 51-53, figs. 24, 25 (appear., habits, etc.); 5th Rept. do., 1873, p. 54 (occur. in N. J. and Md., and larval food-plant).

Bostrichus (Amphicerus) bicaudatus. THOMAS: 6th Rept. Ins. Ill. [1877], p. 123 (descr. of beetle).

Bostrichus bicaudatus. OSBORN: in Trans. Iowa St. Horticul. Soc. for 1879, p. 94, 1880 (brief notice).

Amphicerus bicaudatus. RILEY: in Amer. Entomol., iii, 1880, p. 51, figs. 11, 12.

Amphicerus bicaudatus. SAUNDERS: Ins. Inj. Fruit, 1883, pp. 33-35, figs. 21, 22 (general notice).

This little, and sometimes quite destructive beetle, has long been known in our Western States, especially in Michigan and Illinois, from its more frequent occurrence there, but has not, until recently, been found in the State of New York. It extends southward, at least, into North Carolina, and in the south-west into Texas.

It is a small, cylindrical, dark chestnut brown beetle, black beneath from one-fourth to three-tenths of an inch long, with a small depressed

head, its thorax quite elevated and spinose, and extended in front in two little horns. The wing-covers are also rough, granulated, and in the male they terminate in two short horns, as shown at *b* in the accompanying outline figures, from which feature it derived its specific name, meaning *two-tailed.*

FIG. 28.—The Apple-twig borer, AMPHICERUS BICAUDATUS; *a*, the female; *b*, the male.

The original description by Mr. Say is quite brief, as follows :

A. bicaudatus. Dark reddish-brown; thorax asperous and bicornate before ; a prominent obtuse spine near the tip of the elytra.

Vegetation Attacked.

The insect, as appears from its common name above given, usually attacks the twigs of apple-trees. It is, however, known to occur upon the pear and cherry-tree under like conditions, and has also been found within grapevines, doing far more harm to them than to fruit-trees, if the statement made to the Department of Agriculture and quoted by Mr. Glover (*loc. cit. sup.*), can be relied upon, viz.: "In Iowa, seven to nine-year-old vines are killed from the root up by these insects, and out of fourteen vines eleven were killed " (they were identified by Mr. Glover as *A. bicaudatus* from examples within the vines).

The statement was also made to the Department, that the same beetle had done much damage in Kansas by boring into twigs of young hickory.

Habits of the Insect.

The habits of this little borer are peculiarly interesting, from the fact that it is the mature insect that inflicts the injury by boring the twigs, instead of the grub or larva, which is the usual depreda-tor. Its method of attack is as follows: Just above a bud, at a point from six to twelve inches from the tip of the twig, it gnaws a hole of about one-twelfth of an inch in diameter (the size of its body), and burrowing therein passes downward into the heart of the twig for the dis-tance of an inch or an inch and a half, or occasionally upward, as represented in Fig. 29. This is done soon after the beetle emerges from the pupa state during the months of May or June, for the purpose of feeding and for shel-ter, and not, as would naturally be supposed, to provide a suitable place for the deposit of its eggs. Both of the sexes, it is said, are found within these burrows, and

FIG. 29.— Burrow of AMPHICERUS BI-CAUDATUS in apple-twigs.

always with the head turned from the opening — a reverse position from that which would follow their pupation within the twig and provision for their subsequent escape through the opening.

Life-History.

Of the early stages but little is known. Judging from the range of food-plants which the beetle allows itself, it is probable that the larvæ will prove to have a corresponding range. Thus far it is known only to breed in grapevines, from which it has been reared by Dr. Shimer. Larvæ sent by him to Dr. Packard, which had been taken from grape, were "much the same form as *Lyctus*, but the head is more prominent and also the sides of the body. The anterior half of the body is con-siderably thicker than behind, and the legs are provided with long hairs ; the end of the body is smooth and much rounded. It is 0.30 of an inch long."*

* Is there still room for questioning a grapevine food-plant for this larva ? In 1872, Prof. Riley wrote in reference to it : " The probabilities are that Dr. Packard's description [above quoted] was in reality from this last-named species" [*Sinoxylon basilare*] (4th Mo. Re-port, p. 52). In 1873, Dr. Shimer having communicated the fact that he had found a note attached to an example of *A. bicaudatus*, of its having been bred from grapevine, Prof. Riley concedes that it " substantiates the statement in Packard's *Guide*, giving at least one known food-plant for the larva, and proving the great similarity between it and that of *Sinoxylon basilare* Say" (5th Mo. Report, p. 51). In 1880, we have this statement un-der a notice of *A. bicaudatus :* " The breeding habits of the insect are not yet known with certainty, for while Dr. Henry Shimer found certain larvæ in grapevines which he con-jectured to be of this species, yet they were doubtless those of an allied beetle *Sinoxylon basilare* Say.— C. V. R. in *New York Tribune*" (*American Entomologist*, iii, 1880, p. 51).

In addition to our ignorance of the early stages of this insect — of the eggs — of the larval duration — of the pupation, but little is apparently known of its final perfect stage. And the knowledge that we have of it does not appear to sustain the following inferences :

" We may infer that it comes to maturity late in the summer, and flying into our orchards and vineyards, the beetles bore into twigs during the fall. Here winter overtakes them, and they hibernate in the holes, some of them dying; but most of them surviving until spring, when they continue feeding for a while, and afterward repair to the forest again to propagate their kind " (*loc. cit. sup.*).

This insect is of rare occurrence in the State of New York, and in this region of the United States, and it has, therefore, not been my privilege to study its habits. It appears not to have been carefully studied, for its literature is quite limited, but in all the observations upon it that I have been able to find, I can see nothing to warrant the hypothetical life-history above given. I find no observation of the beetle entering twigs in the autumn, nor any direct testimony of actual *hibernation* in apple-tree or pear-tree twigs. That the beetles have been found within them in the winter is mentioned by different writers, but the only direct statement of personal observation of the fact, published, is that of Mr. Walsh, quoted below, in connection with statements of others bearing upon the burrowing habits. Of the fact that they are to be so found, there can be no doubt, as I have before me a letter received from Prof. Riley, in reply to an inquiry made while writing this notice, in which he asserts : " There is no doubt about *Amphicerus bicaudatus* hibernating within the bored twigs and various trees. I have often received the beetles in the twigs at different times during the winter and spring."

With no desire to be hypercritical, but only to educe the truth, it must be said that neither the above statement from its distinguished source, nor any other known to me, *show* actual hibernation. No mention is made of the condition of the beetle when observed. Although seen alive in midwinter, examination, longer delayed, might have shown that they failed to hibernate. If still alive in early spring — a few dying survivors among many dead companions, as were the examples found by Dr. Shimer in the month of April (see below) — the condition certainly would not be what we understand by "hibernation," nor would a hibernation terminating in death be of the slightest economic importance.

The earliest notice that I find of the boring habits of the beetle is that given by Dr. Harris, in 1854 (*loc. cit.*) He states : " Prof. S. P. Lathrop, of Wisconsin University, and Mr. T. E. Wetmore have sent specimens to me, with accounts of the depredations of the insects, which *are found burrowing in the pith of the branches of the apple-tree, during the spring.*"

Dr. Fitch, in 1856, represents " the twigs withering and *their leaves turning brown in midsummer*," as the result of the borings.

Mr. B. D. Walsh, in 1865, states that the beetle may be found in its burrows in the month of June, also, " I have captured the perfect insects in the woods in September ; and as I once found a single specimen, in the usual situation in an apple-twig, so early in the spring that it must have been there all winter, I infer that they often pass the winter in a perfect state. The great bulk of them, however, bore the apple-twigs in June, and not in the preceding autumn ; and *I have taken several in June when they were only just commencing their holes*, so that half their bodies stuck out in the open air."

The statement in the *American Entomologist*, of December, 1868, p. 80, of Messrs. Walsh and Riley, that " we have found the beetles in them [the burrows] head downwards, in the middle of winter," may, perhaps, be based only upon the above observation of Mr. Walsh.

Dr. Shimer, on April 25, 1868, found many specimens of the beetle, of which the larger proportion were dead, in galleries of dead wood of grapevines. They had doubtless been dead for some time, for he writes of them : " the dead specimens were passably fit for preserving." Upon a subsequent occasion [time not stated] he found two of the beetles boring into grapevines that had been winter-killed.

Prof. Riley, in his *4th Report on the Insects of Missouri*, in 1872, states as follows: " Both the male and female beetles bore these holes, and may always be found in them, head downwards [? turned from the entrance], during the winter and spring months."

Mr. Glover, in the *Report of the Commissioner of Agriculture* for 1872, p. 118, repeats the above statement, but obviously quoting from others.

Dr. Thomas (in 1877) presents reasons for believing that they breed in hickory.

Prof. Osborn, in 1879 (*loc. cit.*) assumes, apparently, that the beetles are in the twigs during the winter, and that the burrows are made for winter protection.

In the *American Entomologist*, for February, 1880, Prof. Riley repeats the winter occupancy of the burrows by the beetles.

Carefully weighing the above, would not a proper conclusion be, that the beetles, as a rule, and perhaps without exception, enter the apple-twigs in or about the early part of the month of June? They have been seen doing so at that time, and at no other. Otherwise this diffi-

17

culty presents itself : The insects " maturing late in summer " (August–September), come forth from their breeding places, are seen abroad " in September," " bore into twigs during the fall " (October–November), " hibernate in the holes " (December–March), " continue feeding for awhile in the spring," thus causing the withering of the leaves (April–June), emerge from their shelter, mate," and repair to the forest again to propagate " (in July). Where, under a calendar thus arranged, would time be found for the three earlier stages of the egg, larva and pupa?

The insufficient and unreliable data that we possess for the completion of the life-history of this insect have been dwelt upon at some length, in the hope that it will be taken up and completed by some of our entomologists or orchardists in the Western States, where the insect is of common occurrence.

We would expect, as the result of careful investigations, to find the insect as a larva within the burrows made by it in grape canes, oak, hickory, or in whatever food-plant it may hereafter be discovered, at any time during the winter and through March and April, perhaps associated in

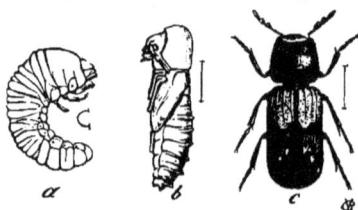

Fig. 30. — The Red-shouldered Sinoxylon, SINOXYLON BASILARE, *a*, the larva ; *b*, the pupa ; *c*, the imago.

the grapevine with its fellow Ptinids *Sinoxylon basilare* and *Lyctus opaculus* (Figs. 30, 31): that passing into its pupal stage from about the middle of May to first of June, the beetle would emerge early in June ; that it would at once find its mate and the female deposit her eggs, in accordance with the law known to control so many of our Coleoptera and other insects — to provide, *first*, for the continuance of the species : that this accomplished, the beetles early in June bore into twigs of apple, pear, etc., as before stated, for food : that their lives within these burrows may be and usually are prolonged for several months, at times, even through the winter and into early spring — one of the beetles having been kept alive for five months under unnatural and, therefore, presumably unfavorable conditions

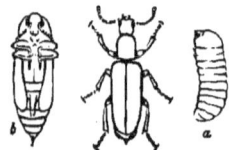

Fig. 31.— LYCTUS OPACULUS: *a*, the larva ; *b*, the pupa ; *c*, the imago—all enlarged (from Packard).

— within a small vial with some grapevine for food (Riley) : that they do not emerge from these burrows, but complete their existence therein — most of them dying in the winter, but some lingering for a few weeks longer.

Such a life-history would seem to meet all the conditions under which the insect has been reported.

Injuries from the Insect.

The injuries reported from this insect have not shown it to be a very dangerous pest. Dr. Fitch simply states of it that it occasions the withering of particular twigs, and their leaves to turn brown through consuming the heart of the twig a few inches in length. — It burrows in the pith of the young branches of the apple-tree during the spring; the branches above the seat of the attack soon die (Harris). — The only damage it occasions is that the bored twig generally breaks off at the bored part with the first high wind (Walsh). — The bored twigs most always break off by the wind, or else the hole catches the water in spring and causes an unsound place in the tree. If the twig does not break off, it withers and the leaves turn brown (Riley).

The amount of harm above stated would not be serious in orchards where the moderate pruning of the last growth could be borne, but in nurseries it would be of much more account; and, in the event of the abundance of the insect, might cause the death of the trees.

The injuries to grapevines, noticed in a preceding page, were more severe, but it is by no means certain that other species were not associated with the *Amphicerus* in that instance, and were instrumental in the death of the vines.

Remedies and Preventive.

Until the early stages of the insect are known, no serviceable remedies can be suggested. It may be both practicable and desirable to employ them against the larvæ when their food-plants shall be discovered. But when the beetle stage is reached, and its operations apparent in the withered twig, if, as we believe, its eggs have already been deposited, no further harm can result from it, and it would be but a waste of time to collect the tunneled twigs for burning, as hitherto recommended.

In a nursery where a strong attack is expected, it would be well to make the effort to prevent the attack. If, at about the time when the beetles make their appearance, the young trees are showered with Paris green in water, the beetle, in attempting to enter the twigs, would probably be poisoned in eating through the bark.

A Beetle of Similar Habits.

In the *American Naturalist* for September, 1872 (page 747), we find the following notice:

We lately received from Mr. Matthew Cooke, of Sacramento, Cal., some pear-twigs in which the above-named beetle [*Polycaon confertus* Leconte] was boring in exactly the same manner as our common apple-twig borer, *Amphicerus bicaudatus.* Mr. Cooke says that the *Polycaon* is quite injurious to apple and pear trees, and also to the grapevine. Thus, from what we know of its natural history, we may safely infer that its habits do not differ essentially from those of *A. bicaudatus, i. e.,* the beetle bores for feeding purposes in living twigs of fruit-trees and grape-vines, never, however, ovipositing in such twigs, and both male and female being concerned in this destructive work. Both species live, in all probability, as larvæ in the dead and dry wood of forest-trees [C. V. Riley.]

In consideration of the above species, *Polycaon confertus,* being so nearly allied in classification with *Amphicerus,* it is interesting to find this close conformity in habits. They are members of the same sub-family of the *Ptinidæ,* viz., *Bostrichinæ.* While the latter, in the recent arrangement of Drs. Leconte and Horn, pertains to Tribe ii (*Bostri-chinæ*), *Polycaon* has place in Tribe iii (*Psoini*), while *Lyctus,* associ-ated with *Amphicerus* in grapevines, follows immediately after in the subfamily of *Lyctinæ.*

It is always gratifying to the economic entomologist to discover identical habits among the different members of a family, as above men-tioned, as it gives the promise and almost the assurance that measures which may be found effectual in controlling one of the species will be equally serviceable against the others.

— — . —

Lema trilineata (Olivier).

The Three-lined Leaf-beetle.

(Ord. COLEOPTERA: Fam. CHRYSOMELIDÆ.)

Crioceris trilineata OLIV.: Encyc. Method., vi, 1790, p. 203, No. 29; Entomologie, vi, p. 739.

Crioceris trilineata. FABR.: Mant. Ins., i, 1781, p. 90, No. 49.

Crioceris trilineata. HARRIS: Ins. Mass., 1841, pp. 95–96 ; Ins. N. Engl., 1852, pp. 104–105 ; Ins. Inj. Veg., 1862. pp. 118–119, f. 53.

Crioceris trivittata SAY: in Journ. Acad. Nat. Sci. Phila., iii, 1856, p. 429.

Lema tririrgata LECONTE : Smithson. Contrib. Knowl., xi, 1859, p. 22.

Crioceris (Lema) trilineata. FITCH : in 10th Rept. Trans. N. Y. St. Agricul. Soc., xxiv, 1864, pp. 441–447.

Lema trilineata. RILEY: in Amer. Entomol., i, 1868, p. 26, figs. 16, 17 ; 1st Rept. Ins. Mo., 1869, pp. 99–100 (from Amer. Ent.).

Lema trilineata. PACKARD ; Guide Stud. Ins., 1869, p. 503, f. 494.

Lema trilineata. REED : in [2d] Rept. Ent. Soc. Ont., 1872, pp. 66–67. figs. 65, 67.

Lema trilineata. CROTCH : Ch. List. Coleop. Amer., 1873, p. 94, No. 5554.

This insect was received from Clyde, N. Y., in the month of July, as quite destructive in potato fields. They were apparently increasing very rapidly from the many clusters of the egg (twenty or more in the cluster, laid upon the leaves.

This species has long been numbered among the many insects injurious to the potato, and is one of those that feed upon it in both its larval and perfect stages. It sometimes occurs in sufficient numbers to be very destructive in New York and the eastern States, while proving comparatively harmless elsewhere. In portions of New York, it has been reported as destroying the potato vines for successive years (*Practical Entomologist*, i, 1866, p. 113). It was quite abundant in the central counties of the State in the year 1864, and, from being regarded as a new enemy, much alarm was excited, from apprehension of its continued increase (Fitch, 10*th Report*, p. 441). In several localities in the Provinces of Ontario and Quebec it is reported as having inflicted an unusual amount of damage among field potatoes (*Report Entomolog. Soc. Ontario for* 1871, p. 67).

The Larva.

The larva has the peculiarity of covering its body with its excrements, its vent being conveniently located for that purpose upon the upper side of its terminal joint, as shown at *b* in Fig. 32. As the material is discharged, it rests upon the surface, while successive discharges force it forward toward the head until the entire back is covered with it, as in the side view of the full-grown larva at *a*, in the figure.

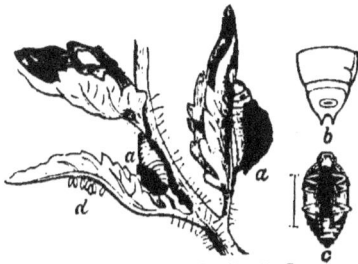

FIG. 32.—The Three-lined Leaf-beetle, LEMA TRI-LINEATA: *a*, the larva; *b*, its terminal joints; *c*, the pupa; *d*, the eggs.

The covering probably serves the purpose of hiding it from its enemies, and perhaps of making it distasteful to birds which would otherwise prey upon it.

In addition to the figure of the larva given, but a few words need be added in description. It is a soft slug-like creature of the general character of the well-known larva of the Colorado potato-beetle. It is transversely wrinkled, with a line of rounded tubercles upon the sides, low down. Its color is dull yellow when young, changing to brighter yellow as it matures. The head is of about the size of the first segment of the body, flattened, smooth and black. The adjoining segment is also black. The body is oval, becoming quite thick behind its middle, and very convex on the back. The three pairs of legs on the first three joints are shining black. Behind these upon other joints are retractile tubercles which, serving as prolegs, aid in locomotion. When . full grown the larva measures three-tenths of an inch in length and about one-half as broad.

The Beetle.

The beetle is represented, in enlargement in Fig. 33. It is about one-fourth of an inch long by one-eighth of an inch broad. Its sides are parallel, with its upper and lower surfaces convex. The antennæ are brownish-black, eleven-jointed, and about half as long as the body. The head is nearly as wide as the thorax and pale yellow; the eyes black and projecting, having behind them a transverse groove. The thorax is as wide as long, constricted in the middle, and marked with two black dots, one on each side of the middle, which are sometimes quite conspicuous.* The wing-covers are yellow with three parallel black stripes, viz.: one at the meeting of the covers, which narrows toward their tip and hardly reaches it, and a broader one almost upon the outer border of each cover. Intermediate to the stripes are several lines of delicate punctures. The legs, which are short, are yellowish-red, and the four-jointed feet are black.

FIG. 33.—The Three-lined Leaf-beetle, LEMA TRI-LINEATA.

Life-History.

The life-history, summarized from the careful and detailed account given by Dr. Fitch, is the following :

The insect hibernates in its beetle state, and probably resorts to forests for shelter.† Coming abroad in May, they feed sparingly upon such vegetation as may be agreeable to them, until about the middle of June when the potato leaves are sufficiently advanced to furnish them with food, and place for the deposit of their eggs. The yellow, oval eggs are laid in clusters of ten or more, usually upon the underside of the leaves. They hatch in about a fortnight. The young larvæ feed in company side by side, either at the tip or on one of the margins of the leaf, but as they approach maturity, they distribute themselves over the leaves. Having attained their growth, they descend from the plants and enter the ground for pupation, where the larva constructs a cell, which it lines with a frothy secretion proceeding from its mouth, in such a quantity as to entitle it to the name of a cocoon.

*A variety occurs in which there is an additional black spot intermediate to the two ordi-nary ones, named var. *tripuncta*, by Fitch. The original description of Fabricius, from ma-terial credited to the Cape of Good Hope, and of Olivier, give three thoracic spots.

†Prof. Riley, in his 1st Missouri Report (as cited), represents the insect as hibernating in the pupa state —" staying in the ground all winter, and only emerging at the beginning of the following June." But Dr. Fitch had found the beetle in a torpid state, on the surface of the ground under boards on the 20th of April.

According to Dr. Harris, the eggs require about two weeks for hatching, and the insect passes about the same length of time in its larval and pupal states, but it does not appear that these periods had been accurately observed.

Dr. Fitch has expressed his belief that there are more than the two annual broods ascribed to it by most writers, viz.: the beetles appearing early in June, and again toward the end of July and early in August. He had noticed their presence in potato fields throughout the season, together with the larvæ in various stages of growth, from which it was " evident that they are not periodical, but are continually coming forth, one after another, and depositing their eggs from the commencement of their operations in June until the chilly nights of autumn suspend their work." This would conform to the continued reproduction of the successive broods of the asparagus beetle, *Crioceris asparagi*, to which *L. trilineata* is closely allied — the twelve United States species of *Lema* of the Crotch Check list uniting with the two of *Crioceris* to form the Chrysomelid tribe of *Criocerini*.

The Insect Sometimes Carnivorous.

From seemingly good authority it appears that this beetle is not an unmitigated pest, but that it renders some compensation for its occasional destructiveness, in preying upon one of its associates and kindred - - a far greater pest. In *Field and Forest* for July, 1877, we find the following :

The three-striped potato-beetle, *Lema trilineata*, has been doing good service this season in the west by destroying the eggs of the Doryphora, or Colorado potato-beetle. Several correspondents in Medina county, Ohio, send specimens of the insect, and report that although the potato-beetles are as plenty as ever, the larvæ are unusually scarce, which is attributed to " the new friend of the farmer." A few were seen last year, but the present season they are very numerous.

As the Lema has always fed upon the foliage of the potato, it is hoped that it will continue and extend its carnivorous habits.

Remedies.

When the insect is not very abundant, but in sufficient force to impair the foliage of the plants, it may be kept within harmless limits by occasional hand-picking, or brushing into pans of water and kerosene oil. This would be more effectual in the larval stage, as the beetle is easily alarmed and readily takes wing and withdraws itself from threatened danger.

Should the insect become very abundant, it may be destroyed by the methods used for the control of the Colorado potato-beetle, viz.: The application of Paris green or London purple.

As most of the many enemies which have acquired the habit of prey-ing upon the Colorado beetle during the several years of its excessive prevalence will, in all probability, as readily feed upon the three-lined leaf-beetle, we need not apprehend, for some time to come, serious ravages from it in our potato fields.

Tribolium ferrugineum (Fabr.).

(Ord. COLEOPTERA: Fam. TENEBRIONIDÆ.)

FABR.: Spec. Ins., 1781, i, p. 324.—STURM: Faun. Deuts., 1807, ii, p. 228, pl. 47, f. d D.— WESTWOOD: Introduc. Class. Ins., 1839, i, p. 319, f. 39, 2.—HORN: Trans. Amer. Philos. Soc., 1871, xiv, p. 365.— MULLER: in Trans. Ent. Soc. Lond., 1873, p. x, Proc. (same in Canad. Ent., 1873, v. p. 156).— BANDI: Deuts. E. Z., 1876, p. 230.— HAGEN: in Proc. Bost. Soc. Nat. Hist.,1878, xx, p. 59.—SCHIÖDTE: Nat. Tidss.,1879, ii, pp. 487, 563, 587, pl. 10, figs. 18–22.— OLLIFF: Ent., 1881, xiv, p. 216.—LUCAS: Ann. Ent. Soc. Fr., 1883, ser. iii, vi, p. lxxi, Bull.

Synonyms.*

    navalis FABR.: Syst. Ent., 1775, p. 56.
    cinnamoneum HERBST: Kaf., 1792, iv, p. 170, pl. 42, f. 8 h H.
    testaceum FABR.: Ent. Syst. Suppl., 1794, p. 179.
    castaneum HERBST: Kafer, 1797, vii, p. 282, pl. 11, f. 3 E.—SCH. Syn. Ins., 1806, i, p. 153.— McLEAY: Ann. Jav., 1825, p. 47.—LUCAS: in Ann. Ent. Soc. Fr., 1854, ser. iii, vol. ii, p. 51; ibid., 1855, ser. iii, vol. iii, p. 249, pl. 13, f. 3.
    ferruginea SAY: Bost. Journ. Nat. Hist., 1835, i, p. 188; Compl. Works, Ed. Lec., 1869, ii, p. 659.— DEJEAN: Cat., 1836, ed. 3, p. 221.
    ochracea KNOCH: Dejean Cat., 1836, p. 221.
    rubens LAP.: Hist. Nat., ii, p. 220.— Dejean's Cat., 1836, ed. 3. p. 221.

Messrs. Durant & Co., grain dealers, of 475 Broadway, Albany, sent for examination, October 25, a package of "middlings" (coarse wheat flour), which had been returned to them as being infested with insects. The material had been received by them about three weeks previous, from a firm in Chicago.

The Insect Swarming in Flour.

Examination of the flour showed it to be literally swarming with *Tribolium ferrugineum* (Fabr.), shown at *a* in Fig. 34. So abundant

---

* The references and synonymy have been kindly furnished by Mr. Samuel Henshaw, of the Boston Society of Natural History.

were they that the amount of the flour that could be lifted upon the point of the large blade of a pocket-knife would contain one or more of the beetles. With the lid of a tin box, an inch and a half in diameter, drawn upward three or four times for four or five inches upon the sides of the bag holding the flour, about five hundred of the beetles were taken. A closer examination of the flour disclosed a few pupæ and a large number of the larvæ. The latter, represented at *b* in the figure, were perhaps one-tenth as numerous as the perfect insect, and were of greatly varying sizes, from apparently full-grown to those about one-twentieth as large. They were quite rapid in their movements, and could hardly be taken from the flour, from the facility they displayed in burying themselves and eluding capture. The beetles were far more active in the evening by gas-light, when they would come to the surface of the flour in the glass vessel in which they were confined, where for some reason, a marked tendency was shown to collect in clusters, clambering over the backs of one another for the purpose. During the several days that they were under my observation, not a single individual was seen to take wing for flight, or to climb up the outwardly sloping side of the glass vessel.

FIG. 34.—TRI-BOLIUM FERRU-GINEUM: *a*, bee-tle; *b*, larva; *c*, anal prolegs—all enlarged. (After West-wood.)

Heat Recommended for Killing it.

Answer was returned the Messrs. Durant, of the name of the insect, its general distribution over the world through commerce, its known habits in Europe as a grain and flour pest, the limited knowledge of it in this country, and recommending an exposure to heat as the best known method of destroying it when infesting grain in bulk or in flour. A moderate degree of heat — from $120°$ to $130°$ Fahr., con-tinued for a few hours, would in all probability suffice to kill all of the eggs, the larvæ and the pupæ in the material submitted, while a higher temperature — perhaps $150°$ or more — would be needed for the beetles.

A Well-known Grain Pest in Europe.

This species has long been known to infest meal, grain and various vegetable stores in Europe, and it is somewhat singular that it has not been brought to popular notice as a pest in this country. I am not aware that any of our entomologists have written of its depredations upon grain products.

Observed as a Museum Pest.

Dr. Hagen has written of injury done by it to the collection of in-sects of the Museum of Comparative Zoölogy at Cambridge.* After

---

* Museum Pests observed in the Entomological Collection at Cambridge.—*Proc. Bost. Soc. Nat. Hist.*, Oct. 23, 1878, xx, p. 56–62.

18

mentioning other species of beetles which had attacked the collection, as *Dermestes lardarius*, *Attagenus megatoma*, *A. pellio*, *Anthrenus varius*, *A. scrophulariæ* and *Ptinus fur*, he states:

Besides these beetles, I made the acquaintance of a pest which I at first entirely underrated, namely, *Tribolium ferrugineum*. The species is cosmopolitan, and as I never heard of damages done by it, I did not at once give it much attention. It was imported several times with large collections of insects from the East Indies. The flat body of the larva, as well as of the beetle, make it particularly fit to enter boxes through the smallest crack. As the collections were exceedingly large, it was impossible to take care of them immediately in a thorough manner, and I observed an alarming increase of the insect. It was rather difficult to overcome, but I succeeded by incessant care, by throwing away the worst infested insects, partly by killing individuals which I forced to come out of the body of specimens by filling the whole box with tobacco smoke. I saw the beetles and larvæ running out when the smoke began to fill the box, then I closed the box for an hour or two, when I found them all dead.

Confirmation of its Carnivorous Habits.

A carnivorous diet must be exceptional to this species and appears hardly to be recognized by our entomologists. A prominent coleopterist who had expressed his belief that it fed entirely upon vegetable material, even after he had been given the reference to the paper of Dr. Hagen, seemed unwilling to admit that it might at times assume carnivorous habits, offering in extenuation of the doubt, the fact that in museums and in the bindings of books, starchy materials are often to be found upon which insects may feed when they have credit for a different diet.

The observations of Dr. Hagen find abundant support in statements made by European writers. The following are some of them, referring to this insect.

Mr. Ingpen has discovered *Stene ferruginea* in bran. "This species is more general in its habits than the preceding [*Tenebrio molitor*], since I have frequently discovered both the imago and exuviæ of the larva in the bodies of old and ill-preserved specimens of exotic insects" (Westwood, *Introduction Classif. Ins.*, i, p. 319).

In opening some paper envelopes of Lepidoptera received from Abyssinia, many of the specimens were eaten and had gone to pieces through the attack of *T. castaneum* [= *T. ferrugineum*]. The larva, pupa and imago were found and figured (Lucas, *Ann. Soc. Ent. Fr.*, 1855, ser. 3, vol. iii, p. 249).

The beetle and larva occur in old bread, when meal or flour are imported, and in stores of natural history specimens (Taschenberg, *Prakt. Insectenkund*, 1879, ii, p. 90).

Ordinarily found in meal and rice, but it attacks even collections of insects (Lacordaire, *Coleopt.*, v, p. 323).

Other Notices of the Insect

Mr. W. E. Saunders also makes mention of its carnivorous habits in
Canada. He states that it infests patent food and similar substances,
and that it eats the dead bodies of other beetles. (Insects Injurious to
Drugs — *Canadian Entomologist*, May, 1883, xv, p. 82.)

Dr. Horn, in his Revision of the Tenebrionidæ of America, North of
Mexico, *Trans. Amer. Philosoph. Soc.*, 1871, xiv, N. S., p. 365, merely
groups it with its only known congener in our fauna, *C. madens* (Charp.),
remarking of it that the former species is ferruginous, the latter black,
and that in length, they vary from .16 to .20 inch, *C. madens* being the
larger. Both are found abundantly wherever meal or grain is stored.

It is named in the list of Drs. LeConte, Horn, and Leidy, of " Insects Introduced by the Centennial Exhibition at Philadelphia," as
occurring in mouldy specimens of straw goods from Italy.

I find no other notice of the species by our American authors.

The following interesting statement of its infesting a cargo of pea-
nuts was communicated to the Entomological Society of London, by
Mr. Albert Muller:

In the summer of 1863, a cargo of ground-nuts (*Arachis hypogœa*)
arrived in the port of London, direct from Sierra Leone. On arrival
the usual samples were drawn, when it turned out that the husks were
riddled by countless holes, while the kernels were half eaten up by
myriads of larvæ and imagines of *Tribolium ferrugineum*. So completely had they done their noisome work, that in the numerous samples examined scarcely an intact kernel could be found. If a nut was
opened the whole interior was often found to be converted into a living
conglomerate of larvæ, pupæ and imagines of *Tribolium*, accompanied
by the larvæ and perfect insects of a *Rhizophagus* preying on the former,
the whole mass being wrapped up in a layer of cast-skins and excrement.

Brachytarsus variegatus (Say).

(Ord. COLEOPTERA: Fam. ANTHRIBIDÆ.)

SAY: in Journ. Acad. Nat. Sci. Phila., v, pt. ii, 1827, p. 251 (*Anthribus*); Complete Works, Ed. Lec., 1869, ii, p. 314.

WALSH: in Journ. Ill. St. Agricul. Soc., 1862, pp. 8–12, figs.; in Proceed. Bost.
Soc. Nat. Hist., 1864, ix, p. 309.

LeCONTE: in Proceed. Amer. Philosoph. Soc., xv, 1876, p. 406.

Synonyms:

> *sticticus* GYLL.: Schön. Curc., 1833, i, p. 172 (Brachytarsus).
> *obsoletus* FAHRÆUS: Schön. Curc., 1839, v, p. 107.

Examples of this little Ryhnchoporid beetle were sent to me under date of August 28, by Mr. C. A. Gillett, of Shortsville, Ontario county, N. Y., with this communication :

Depredations upon Wheat.

I send you with this mail a few specimens of an insect that I find in a bin of newly-threshed wheat. I also send a few kernels of wheat that were taken from the same bin. They show the work of an insect, presumably this one, in its growth from a larval state. Does it not seem that wheat is yet to be affected by this insect much as the pea is by the pea-bug ? The few kernels inclosed with the bugs were thought to be perfect when put in the vial. If any marks are detected upon them, it must be owing to the eating of them by the bugs.

The sound kernels were found to have been eaten into when received; and during the following days that they were observed, the cavities increased in number and in size. There could, therefore, be but little doubt but that the burrowing of the wheat in the bin had been by this insect. The kernels taken from the bin showed different degrees of injury, from small rounded holes on the outer surface, to the excavation of nearly all of the interior.

It is not probable, however, that the fears expressed of serious injury to wheat by this insect are to be realized. It belongs to the Rhynchopora, or snout-beetles, and to the family of *Anthribidæ*, named from the Greek *anthos*, a flower, and *tribo*, to destroy. Most of the larvæ of this family find their food within the seeds and stems of plants.

The Larvæ of Brachytarsus Parasitic.

Dr. LeConte, in his " Rhynchophora of North America," Introduction, page xiv, remarks: While the food of the Rhynchophora is almost universally vegetable tissues, either living or dead, *Brachytarsus* is a parasite upon a Hemipteron, of the genus *Coccus*, as narrated by Nordlinger, *Stettin Ent. Zeitung*, 1848, p. 230; Lacord., *Gen. Col.*, vii, 481.

A species occurring in Europe, the *B. varius* Fabr., is known to live upon a *Coccus* infesting pine trees.

See also Prof. Westwood's remarks upon the parasitical connection of some of the species of this genus with the *Coccidæ*, in his Classification of Insects, i, p. 332.

The larvæ of this genus *Brachytarsus*, according to Dr. Le Baron, are found under the scales of bark-lice, where they are believed to be parasitic.

B. variegatus not a Wheat Pest.

It may, therefore, be inferred that the attack upon the wheat above noticed was not by the larva, and that the presence and feeding of the beetle was but accidental. It is known to be quite a general feeder, finding its food in various vegetable substances with which it may chance to be associated. Hence for the present, it will not be necessary to number *Brachytarsus variegatus* among the insect enemies of the wheat.

The attack was not a serious one. Upon inquiring of its character, Mr. Gillett informed me, that in looking over the wheat as it lay upon the floor, a perforated kernel could be found in an area the size of his hand, and more than twice as many of the beetles in the same space.

Description of the Beetle.

Appended is the original description of the insect as given by Mr. Say, together with his observation of its occurrence upon the smut of wheat, upon which it had probably fed.

A. variegatus. Varied, with blackish and dull yellowish; elytra, each with two larger spots.
Anthribus variegatus, Melsh. Catal.
Body dull ochreous, varied with blackish, with very short hair; *head* plane. dusky, paler toward the tip; antennæ moderate, pale rufous, three terminal joints fuscous; *thorax* much varied with blackish, which does not extend on the anterior margin; *elytra* with hardly obvious striæ, with many small orbicular, blackish spots, and two larger spots on each, of which one is near the sutural base, and the other rather beyond the middle, near the suture; *feet* pale rufous; thighs blackish in the middle.
Length more than $\frac{1}{10}$ of an inch.
The two spots of each elytron are sometimes united by an intervening blackish sutural line.
The species is not uncommon, and I have found it on the "smut" of wheat. Mr. Lea took eighty individuals from six heads of wheat.

Distribution.

This insect is generally distributed throughout the Atlantic States, associated with *B. limbatus* Say, from which the variety (*obsoletus*) in which the elytral spots are obsolete, may be distinguished by its more slender form.

Aramigus Fulleri (Horn).

Fuller's Rose Beetle.

(Ord. COLEOPTERA: Fam. OTIORHYNCHIDÆ.)

HORN: in Proc. Amer. Philosoph. Soc., 1876, xv, p. 94 (orig. description); in Canad. Entom., xvi, 1884, p. 184 (remarks on distribution).

RILEY: in Ann. Rept. Commis. Agricul. for 1878, pp. 255–7, pl. 7, f. 2; Id., Rept. Entomol., 1879, pp. 50–1, pl. 7, fig. 2*a–h;* in Amer. Entomol., 1880, iii, p. 26 (occurrence in California); in Rept. Commis Agricul. for 1884, p. 414 (habits in Mass.).

COMSTOCK: in Ann. Rept. Commis. Agricul. for 1879, pp. 250–1 (distribution).

AUSTIN: Check List Coleop. N. A., 1880, p. 44, No. 8841.

LINTNER: in Count. Gent., xlix, 1884, p. 49 (remedies, etc.).

MOFFAT: in Canad. Entomol., xvi, 1884, p. 216 (occurrence in Canada).

During the month of December, last, this beetle was reported to me as occurring in great number in the extensive rose-growing establishment of Messrs. Fricker & Clarke, of Poughkeepsie, N. Y. It was proving very troublesome and destructive in their houses, and in the latter part of November, the beetles could be seen collected in large clusters upon the bushes. Many had been gathered and destroyed, but they still continued to abound.

That they should have been permitted to multiply to this extent was certainly inexcusable neglect.

A Green-house Pest.

This insect, during the last few years, has attracted considerable attention from the injuries committed by it in green-houses. It was first brought to my notice in the year 1874, by Mr. A. F. Chatfield, florist, of Albany, who had found the beetles eating the leaves of camellias in his conservatories and injuring the foliage. Soon after this, it appears to have been discovered in the same nefarious work in the green-houses of Mr. A. S. Fuller and other gentlemen in New Jersey. Examples of it were sent to Dr. G. H. Horn, of Philadelphia, who, in 1876, finding it to be an undescribed species and quite different from any known form, established a new genus for it, and gave it the specific name of *Fulleri*, after the gentleman who had first brought it to his notice, and who is also "a popular author on horticultural and natural history subjects, and well and widely known for his interest in entomology," especially in the Coleoptera.

Distribution and Food-plants.

Since that time, it has been reported from various localities in the

United States from the seaboard to the Pacific ocean, and from Canada. Its greatest injury is committed upon roses grown under glass, by the larvæ feeding upon the tender rootlets — at first merely checking their growth, but finally, when their numbers have increased, destroying the plant. It has also been observed upon the roots of *Geranium* and *Hibiscus*, and in California, is reported as "very destructive to *Dracœnas* (and palms lightly), oranges, cape jessamine [*Gardenia*], and *Achyranthes*, in the order named." In Brantford, Canada, it has been found upon *Abutilon* and *Plumbago* in hot-houses. From Massachusetts, it is reported upon the *Azalea*, "Cissus," and "inch-plant."

Transformations of the Insect.

In the report of the Commissioner of Agriculture for 1878, the insect has been excellently figured in its four stages, and of the first three, scientific description is given. (Fig.35 is from an electrotype obtained from the Department.) From the accompanying account by Prof. Riley, it appears that the eggs are laid in clusters of from ten to sixty, arranged in rows, which are artfully secreted between the bark and the trunk of the plant near its base, or occasionally between the ground and the trunk. Upon the hatching of the eggs (*e*) in about a month, the larvæ (*a*) burrow and begin their feeding upon the rootlets. When fully grown they change to pupæ (*b*) within the ground from which

FIG. 35 — ARAMIGUS FULLERI Horn: *a*, the larva; *b*, the pupa; *c*, the beetle, in side view, and *d*, a dorsal view, with the natural size shown between; *e*, the eggs enlarged and in natural size; *f*, left maxilla with palpus; *g*, under side of head of larva; *h*, upper side of the same.

the perfect insects (*c*, *d*) emerge to pair, deposit their eggs for another brood, and to feed upon the leaves. Upon the under side of the foliage and the branches they may be found at rest during the day, as they are nocturnal in their habits, and only leave their concealment after dark for feeding.

Description of the Beetle.

The following is the description of the beetle, as given by Dr. Horn, in *The Rhynchophora of America, North of Mexico*, loc. cit.:

Form oblong oval, surface not densely clothed with dark brown scales. Head and rostrum longer than the thorax, densely punctured.

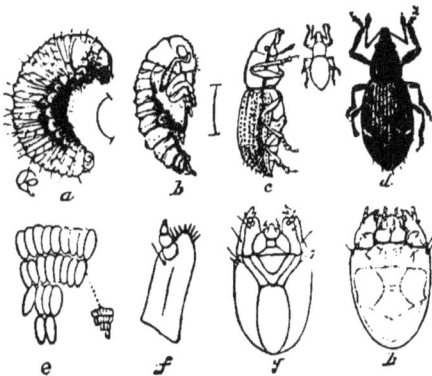

sparsely scaly. Rostrum with feeble ridge on each side from the tip nearly to the eyes. Thorax cylindrical, apex and base equal and truncate, very slightly wider than long, sides feebly arcuate, a fine median line, disc moderately convex, densely punctured, sparsely scaly. Elytra regularly oval, humeri entirely obliterated, base sub-truncate, surface indistinctly striate, and with rows of large, moderately closely placed punctures, intervals flat, not densely scaly, and with very minute sub-erect hairs ; scales dark brown, a whitish or paler stripe beginning at the humerus, passing along the lateral margin, ending in a short oblique fascia at the middle of the elytra. Body beneath sparsely scaly. Legs with scale-like hairs. Anterior tibiæ rather strongly denticulate within, articular surfaces of middle tibiæ not ascendant. Length. 0.26 inch; 6.5 mm.

Remedies.

The best method by which to meet the depredations of this insect, so far as known at the present, is to hunt for the beetles upon their food-plants, and to destroy them. If this be persistently done, the evil can be arrested. Mr. Chatfield informs me that by diligently searching for and killing all that he could find, for two or three successive years, he believes that he has exterminated them from his plant-houses, as he has not noticed them for the past two or three years.

Various experiments have been made with a view of killing the larvæ while preying upon the roots, but the opinion seems to prevail, as the result of such efforts, that it is only to be done at the sacrifice of the plant. A rose-bush which is known to be badly attacked at the roots had better be at once taken up and burned, and the soil that contained it treated with some caustic substance or with heat, to destroy such larvæ as may have been left behind.

If the eggs — which are described as smooth, soft, of a clear yellow color, and elongate-elliptical in shape — are carefully looked for in the places where they are usually hidden, they may be found and readily destroyed, and the injuries of the insect materially checked. Prof. Riley has recommended placing traps for the eggs, of pieces of paper or rags, wound about the trunks, or upon sticks thrust in the ground near them, to be collected at intervals of not exceeding three weeks, for the destruction of the eggs that may have been deposited upon them, by burning, or, if it is desired to use the traps again, by dipping them into hot water.

Cosmopepla carnifex (Fabr.).

(Ord. HEMIPTERA: Subord. HETEROPTERA: Fam. CYDNIDÆ.)

Cimex carnifex FABR.: Ent. Syst. Suppl., 1794, p. 535, No. 162.
Eysarcoris carnifex HAHN: Wanz. Ins. ii, p. 117, f. 198.

Pentatoma carnifex KIRBY: Faun. Bor.-Amer. iv, 1837, p. 275, No. 1.
Cosmopepla carnifex STAL: Enum. Hemipt., ii, p. 19, No. 1.
C. carnifex. UHLER: in Bull U. S. G.-G. Surv. Terr., i, 1876, p. 18; id., iii, 1877,
      p. 402 (localities).
C. carnifex. GLOVER: MS. Notes Journ.—Hemipt., 1876, p. 35, pl. ii, f. 6.
C. carnifex. LINTNER: in Count. Gent., xxxix, 1874, p. 488, c. 1.

This insect is one of the true bugs, belonging, as above recorded, to
the Heteropterous division of the Hemiptera, in which the wings instead
of being deflexed are horizontal, the upper pair, or wing-covers, as they
might more appropriately be called, are thickened or "leathery," with
their tips thin and overlapping. The under wings are thin and mem-
branous. The proboscis, rostrum, or sucker is much like that of the
other Hemiptera, except that it springs from the front of the under side
of the head instead of beneath it, posteriorly near the base of the fore
legs.

Description.

The Imago. — The *C. carnifex* has much the general appearance of
the Harlequin Cabbage-bug, *Murgantia histrionica*,
noticed in my preceding report, having the same colors, but being a smaller insect, and proportionately
broader in form. Its length is about 0.23 of an inch,
and its breadth 0.17 in. The general color is shining
black. The head, thorax and coriaceous portion of the
wing-cases are granulated. The thorax is crossed by
a transverse elevated ridge, marked with dull orange
and is bisected by a slender mesial line of the same
color. Coriaceous portion of wing-covers margined
with orange, which is broader basally, thence becoming
obsolete; margin of abdomen, also, orange. The scutellum is long, pointed, extending over two-thirds of
the abdomen, and is marked with two triangular orange
spots near the tip, one on each side. The antennæ, legs
and proboscis are black. The insect enlarged from the natural size
shown by the lines beside it, is represented at *b* in Fig. 36.

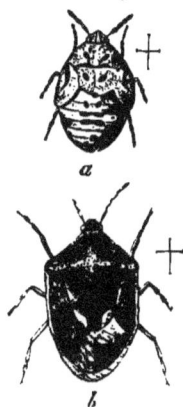

FIG. 36.— COSMOPEPLA
CARNIFEX; *a,* the pupa; *b*
the imago.

The Pupa. — The pupa, shown at *a*, is 0.20 by 0.16 of an inch in
length and breadth. It is dull yellow, with the eyes, antennæ and legs
(except their basal portion) black. The proboscis is black at the tip.
The head has two longitudinal mesial black lines which diverge posteriorly. The thorax has a black line on its hinder margin, centrally,
which is sometimes bisected at the middle, and also two rather large
black spots, centrally on each side pointing backward and sometimes

19

connected with the black marginal line. A rounded black spot rests centrally on each side of the scutellum, and there is also a black dot near each anterior angle. The wing-pads are edged with black behind, and in addition they bear an S-like character in black, extending back centrally from the base. Upon the abdominal segments dorsally are two pairs of shining black points, of which the hinder pair are usually extended laterally into a line, and the front pair are sometimes connected; laterally, are four black dots marking the segments, which are also seen from beneath,

Not Hitherto Known as Injurious.

We find no record of injuries inflicted by this species upon any of our crops, nor notice of its habits by any of our economic writers. Mr. Glover, in his mention of it in his *Manuscript Notes from my Journal*, merely states in referring to a figure that he gives of it, "Insect probably destroys other insects." This supposition may have been drawn from the specific name which it bears, meaning "assassin," "executioner," "villain," or from the known carnivorous habits of *Podisus spinosus* (Dallas), *Perillus circumcinctus* (Stal), *Stiretrus anchorago* (Fabr.), and others, forgetting the very different habits displayed by the members of this large family — that the well-known cabbage pest, *Murgantia histrionica*, is a near relative of *C. carnifex*, belonging to the same sub-family of *Pentatomina*, and that another member of the same, *Euschistes variolarius* Pal. Beauv., feeds upon plants and animals interchangeably.

The Insect Attacks the Potato.

The insect in its larval, pupal and perfect stages, has been sent to me from Souyea, Livingston county, N. Y., about the middle of July, with the statement that they were injuring the vines, seriously, and that they were believed to be poisonous, as the vines, shortly after they had been punctured by the insect, withered and died. Paris green had been tried for killing, but to no purpose.

It also Attacks the Currant.

Subsequently, the mature insect has been sent to me for identification, by Prof. D. P. Penhallow, of McGill University of Montreal, Canada, as an injurious currant insect. It had appeared in large numbers in 1884, attacking the fruit, causing it to fall, and seriously injuring the crop.*

---

* It reappeared in still larger numbers the following summer.

Remedies.

As all the bugs belonging to the order of Hemiptera draw their food from the interior of the plant, sucking the sap by means of a proboscis thrust into the plant, it is useless to attempt to poison them by applications to the surface of the leaves. Many of them are so conspicuous in appearance, of so large size, and occurring often in not remarkable numbers, that their injuries to many of our low crops can be prevented by picking them off by hand, or by beating them from their food-plants by the aid of a short stick into a broad basin of water and kerosene.

When too abundant for this method, or in situations where they cannot be conveniently reached by hand, they will not be able to withstand thorough sprinkling with kerosene emulsions, so applied as to reach their breathing pores — a difficult matter perhaps in some of the species, which are so well protected by their close-fitting and broad wing-covers.

Its Distribution.

This species has quite a broad distribution over the United States. Mr. Uhler records it from Maine to Georgia, Texas, Indian Territory, Nebraska, Kansas, Missouri and Washington Territory; also from Port Neuf, Canada, and from Nova Scotia.

I have found it in large numbers, in former years, at Schoharie, N. Y., in the early part of July, upon some low plants (weeds?) — the species not recorded.

Is Probably not Carnivorous.

In the hope that the insect might, as a redeeming trait in its character, be induced to imitate the habits of some of its relatives, I made request of the gentleman from whom the specimens had been received, to confine them, after depriving them for a time of their vegetable food, with the larvæ of the Colorado potato beetle. The experiment was made, but as they showed no disposition to change their diet, during their trial of two days, it may be inferred that they are strictly vegetarian in their feeding.

Another Attack upon the Potato.

A singular attack upon the potato tuber may properly be recorded here, in connection with the above notice of a new attack upon the stems and leaves. It was brought to my notice several years ago (in 1877), by Mr. Miller, of the Austrian Commission at the Centennial Exposition at Philadelphia.

During his sojourn in the city of New York, he observed in many
of the boiled potatoes brought upon the tables of the hotels and restau-
rants, upon being crushed by the knife, the remains of a channel or
channels leading from the center to the exterior, which may have been
the one-twentieth of an inch in diameter. They were apparently of
uniform size, and often gave off two or three branches extending to the
surface. From his imperfect use of the English language, I cannot
assert positively that I quote him correctly in saying that the walls of the
channels had such consistency as to permit of their cohesion after having
been crushed so as to show their form and branching structure. The
passages, he believed, in all cases, originated at the center, in a black
spot. He regretted that his examinations had not been sufficiently
minute to enable him to state with certainty whether the channels were
entirely uniform in diameter, or if they enlarged somewhat with their
outward extension, as is usual in the burrows of beetles made during
their growth, and by interior-feeding lepidopterous larvæ. No animal
remains were discovered within the tubes, the habit of the borer being,
as was stated to him, to leave the potato while in the ground. The
potatoes affected in this manner had been received from California, while
in those coming from elsewhere, nothing of the kind had been noticed.

The gentleman believed it to be an insect attack, and from the evidence he gave of the ability to judge correctly in such a matter, it is
probable that he was correct. But of the character of the insect, the
time and the method of the introduction of the egg, no opinion of value
can be offered.

Blissus leucopterus (Say).

The Chinch-Bug.

(Ord. HEMIPTERA: Subord. HETEROPTERA; Fam. LYGÆIDÆ.)

Lygæus leucopterus SAY: Heterop.-Hemip. N. A., 1831, p. 14; in Trans. N. Y. St.
    Agricul. Soc. (for 1857), xvii, 1858, p. 774.
Rhyparochromus devastator LE BARON: in Prairie Farmer, v, 1845, p. 287; id., ix,
    1850, pp. 280-1.
Rhyparochromus leucopterus HARRIS: Treat. Ins. New Eng., 1852, pp. 172-3; Ins.
    Inj. Veg. 1862, pp. 197-199, f. 84.
Micropus leucopterus FITCH: in Trans. N. Y. St. Agr. Soc. for 1855, xv, 1856, pp.
    509-529; 1st and 2d Rept. Ins. N. Y., 1856, pp. 277-297, pl. 4, figs. 2, 2a.
The Chinch bug, WALSH: in Trans. Ill. Agr. Soc. for 1859-60, iv, pp. 346-349, p.
    436; in Pract. Entomol, i, 1866, p. 95 (barracading by tarred boards); ib.,
    ii, 1866, p. 21 (in Canada).
Micropus (Lygæus) leucopterus, SHIMER: in Proc. Acad. Nat. Sci. Ph. for May,
    1867, [xix], pp. 75-88 (injuries, habits, destroyed by an epidemic).

Micropus leucopterus. WLSH.-RIL.: in Amer. Ent., i, 1869, pp. 169-171, 194-199.

Rhyparochromus leucopterus. PACKARD: Guide Stud. Ins., 1869, pp. 543-4, f. 547; (as *Blissus leuc.*) in Ninth Rept. U. S. G.-G. Surv. Terr. for 1875, 1877, pp. 697-699, fig.; First Rept. Ins. Mass., 1871. p. 4.

Micropus leucopterus. BETHUNE: in Rept. Ent. Soc. Ont. for 1871, pp. 55-57 (general notice, and presence in Canada).

Micropus leucopterus. LE BARON: Second Rept. Ins. Ill., 1872, pp. 142-156.

Blissus leucopterus. THOMAS: Seventh Rept. Ins Ill., 1878. pp. 40-71; Bull. No. 5, U. S. Entomolog. Commis., 1879, pp. 44, figs. 10, and 1 map; in Amer. Entomol., iii, 1880, pp. 240-242 (influence of meteorological conditions).

Micropus leucopterus. RILEY: Seventh Rept. Ins. Mo., 1875, pp. 19-50, figs. 2-11; in Amer. Nat., xv. 1881, p. 820 (notice of a chinch-bug convention in Kansas); in Amer. Agricul., xl, 1881. p. 476, figs. 1-3; ibid., p. 515, figs. 1-4; in Science, ii, 1883, p. 620; in Rept. Commis. Agricul. for 1884, pp. 403-405.

Micropus leucopterus. GLOVER: MS. Notes Journ.--Hemipt., 1876, p. 489, pl. 2, figs. 16, 17.

Blissus leucopterus. UHLER: in Bull. U. S. G.-G. Surv. Terr., i, 1876, p. 306; in separate, as List of Hemip. West of Miss. Riv., p. 40 (distribution in U. S.)

Blissus leucopterus. FORBES: Twelfth Rept. Ins. Ill., 1883, pp. 32-63 (life-history, nat. enemies, parasites, remedies, etc.); in Bull. No. 2. Div. Entomol., U. S. Dept. Agricul., 1883, pp. 23-25 (experiments with kerosene emulsions).

Blissus leucopterus. LINTNER: in Albany Argus of October 10, 1883; in Count. Gent. for October 18, 1883, p. 841; in Science, October 19, 1883, ii, p. 540; Circular No. 1, N. Y. St. Mus. Nat. Hist., October 18, 1883, 3 pp.; in Thirty-seventh Ann. Rept. N. Y. St. Mus. Nat. Hist. [November, 1884], pp. 53-60.

During the last week in September of 1883, a package of insects in roots of grass was sent to the New York State Agricultural Society, by Mr. M. H. Smith, of Redwood, Jefferson county, New York, with the following statement in regard to them :

I herewith transmit specimens of (to us) a new and formidable grass-destroying insect, together with portions of grass destroyed by them, and also some of the soil, for the purpose of examination. If the insect is known to you, and there is any known way to exterminate it, please inform us at once. The evidence of its destructive work was first discovered in June of 1882, by Mr. H. C. King, of Hammond, St. Lawrence county. At haying time, about the middle of July, he noticed about three acres of his timothy grass to be apparently prematurely ripened. In the fall he observed that there was no aftergrowth, and that the stubble was as dead as if it had been boiled. Search was made among the dead roots without any discovery. The following spring the field was entirely barren of timothy, but some clover seeds and thistles occupied the ground where at least one and one-half tons of timothy to the acre, under favorable circumstances, would have been cut. In June of 1883, Mr. King discovered other fields to be affected in the same manner, and instituted a search which has recently resulted

in the discovery of myriads of the insect, not in the dead grass, but at
the edge of the live grass, where they may be scraped up by handfuls.
They have destroyed about fifteen acres for Mr. King and several acres
for each of several other farmers of his vicinity. They are causing
extreme alarm, and if you can give any relief from this calamity it will
be gratefully appreciated. This is an important grazing locality. In
addition to the timothy, June grass and wire grass are also destroyed.

The Insect Identified.

The insects being submitted to me by Secretary Harison, of the
State Agricultural Society, they were at once, greatly to my surprise,
recognized as the notorious chinch-bug of the Southern and Western
States. It was the first instance of a New York specimen of the species
coming under my observation, nor had I knowledge of its occurrence
within the State, beyond the record of Dr. Fitch of his having met with
three individuals of it. Dr. Harris had seen one specimen in Massachusetts. In each of the above instances, the occurrence was deemed
of such interest and importance that the date of observation was given.*

The Insect Described.

It belongs to the order of Hemiptera, which comprises all of the bugs
proper. It is, therefore, without biting jaws, but takes its food by suction
through a four-jointed proboscis, which, when not in use, is bent beneath
the body. Its size seems quite disproportioned to its destructive powers,
being but three-twentieths of an inch long and one-third of its length
broad. Its body is black and slightly hairy (a fine grayish down) under
a magnifier. The wing-covers, resting flat upon its back, are white with

a subtriangular black spot in the middle of the outer
margin of each, and a few black veins upon their
middle; the feet, claws and enlarged ends of the
antennæ are black, while elsewhere the latter and
the legs are dull yellow. Fig. 37 represents the
insect.

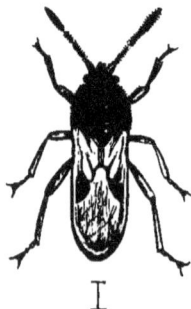

Although the injuries of this insect in wheat-fields
in North Carolina were known over a hundred years
ago, it was not until fifty years thereafter, that it
received from Mr. Say, of New Harmony, Ind., a
scientific name and description. It was referred by
him to the genus *Lygæus*, and given the specific

FIG. 37— The Chinch-bug,
BLISSUS LEUCOPTERUS.

name of *leucopterus*, meaning white-winged. The original description
is as follows :

Blackish, hemelytra white with a black spot. Inhabits Virginia.
Body long, blackish with numerous hairs; antennæ, rather short hairs;
second joint yellowish, longer than the third, ultimate joint longer than
the second, thickest ; thorax tinged with cinerous before, with the basal
edge piceous; hemelytra white, with a blackish oval spot on the lateral
middle; rostrum and feet honey-yellow; thighs a little dilated. Length
less than three-twentieths of an inch.

History of the Insect.

The very serious nature and the extent of the injuries resulting from
this diminutive pest have drawn close attention and study to it, as appears in the many pages devoted to its history, transformations, habits,
etc., by most of our principal writers in economic entomology. It has
been properly characterized as " unquestionably one of the most pernicious insects which we have in the United States ; the locusts of Utah
and California being the only creatures of this class which exist within
the bounds of our national domain, whose multiplication causes more
sweeping destruction than does that of this diminutive and seemingly
insignificant insect " (Fitch in 1855). It is but natural, therefore,
that its literature should prove extensive, and that all that relates to its
past history—its life-stages, habits and present status—means available for
the control of its ravages, and the probabilities for the future, should be
of deep interest to the agriculturist who is at all conversant with its destructive capabilities.

Dr. Fitch, in his second report, has devoted ten pages to its early history in this country. He states that appearing at the close of our revo-
lutionary struggle, about the year 1783, in the interior of North Carolina, it was at first regarded as the Hessian fly, which at the same time
was proving so destructive to wheat-fields on Long Island and in New
Jersey. The insects continued to increase throughout North and
South Carolina and Virginia for several years. As early as 1785 the
wheat-fields in North Carolina were threatened with entire destruction.
The ravages continued for several years thereafter, and the cultivation
of wheat was temporarily abandoned.

In 1809 they again became so destructive in North Carolina that wheat
was not sown for two years. But little was heard of them for a number
of years thereafter, until in 1839, when they again became excessively
numerous in Virginia and North Carolina, and extended their depredations to corn, oats and other grains. A writer states of them: " I
have seen some of my corn so perfectly black with them for two feet
up that no particle of grain was to be seen, but five or six inches of the

tips of the leaves ; and they hung to the under parts of them in knots like little swarms of bees. It takes only one or two days to destroy the corn." In several instances entire fields were burned as the only known method of subduing the pest. The total destruction of the crops appeared inevitable, but most opportunely the season of 1840 proved to be an exceedingly wet one. The insect could not endure the rains, but died in myriads, and the attack was arrested.

At about this time the chinch-bug was noticed along the Upper Mississippi and in northern Illinois. Appearing simultaneously with the advent of the Mormons at Nauvoo (1840–1844), it was believed by many ignorant people to have been introduced by them, and the bugs were accordingly known as " Mormon lice."

It soon extended itself throughout all parts of Illinois and in· adjoining portions of Indiana and Wisconsin. At this time, in Illinois in the middle of extensive prairies, upon parting the grass in search of other insects, Dr. Fitch found the ground swarming and covered with the bugs.

According to other writers, the insect appeared in 1847, in Iowa. In Indiana and Wisconsin, it was first observed in 1854 and 1855. In 1864, the crops in Iowa and adjoining portions of Illinois suffered severely from its attacks. In 1871 it spread in vast numbers over most of the North-western States, and again in 1874 it proved very destructive. In this latter year it was very injurious in Missouri, although its depredations had been serious in certain sections of the State in the years 1854–1857; its first recorded appearance there was in 1836.

The Common Name of the Insect.

Inquiry is not infrequently made of the origin and meaning of the popular name of this insect. Dr. Fitch has given this explanation: At first it was only the wingless larvæ which were supposed to be the depredators, they not having been associated with the mature form with its white wings. The larvæ have a close resemblance to the common bed-bug (*Acanthia lectularia*) formerly known as *Cimex lectularia*. Throughout the Southern States, the bed-bug is everywhere known by its Spanish name of *chincho*, and, therefore, when it was ascertained that the wheat depredator was not the Hessian fly, it came to be distinguished by the name of the *chinch*-bug.

Webster derives the Spanish *chincho* from the Latin cimex, and defines chinch, 1st, as the " bed-bug," and 2d, as " an insect or bug, resembling the bed-bug in its disgusting odor, which is very destructive to wheat and other grains; also called *chintz, chinch-bug, chink-bug.*"

Life-History.

Transformations. — The eggs are laid in the autumn upon the crown or the roots of the plants, where they remain throughout the winter and until the ground becomes warm enough in the following spring for their hatching. Or they may be laid in the spring by the bugs that were the later to mature and have hibernated in the perfect or in the pupal stage.* Occasionally clusters of the eggs are to be found above ground attached to the blades of their food-plants.

The eggs are so minute as hardly to be visible without a magnifier, measuring only about three-hundredths of an inch in length by one-fourth as broad. They are of a dull reddish color, elongate oval with the exception of being flattened at one end, where they bear four small tubercles, as at *a* and *b* in Fig. 38. About five hundred eggs are deposited by each female, at intervals, extending over several weeks. They hatch in about two weeks, when the larva appears at first of a pale yellowish color, but soon changes to red, except the two anterior segments of the body, and the legs, which are yellowish. In this stage it appears as shown in *c*. After the first molt, it becomes bright red, with a pale band across the middle of the body. After the second molt, the wing-pads begin to make their appearance, as at *f*, and the general color has become darker, with the pale band still conspicuous. Another molt brings it to its pupal stage, as at *g*, with distinct wing-pads; its anterior portions dark brown (fuscous), and its abdominal portions grayish, except the tip which is brown. The entire period required (when not interrupted by hibernation) for the transformations from the egg to the perfect insect, is from five to seven weeks.

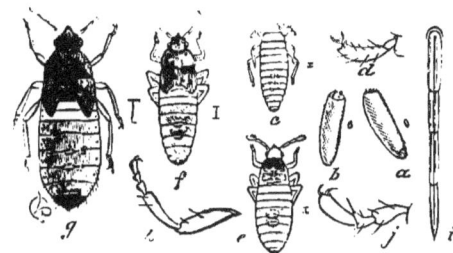

FIG. 38. — EARLY STAGES OF THE CHINCH-BUG; *a, b,* eggs; *c,* newly-hatched larva; *d,* its tarsus; *e,* larva after first molt; *f,* same after second molt; *g,* the pupa; *h,* leg of mature bug enlarged; *j,* tarsus of same, more enlarged; *i,* proboscis or beak enlarged.

The early brood. — During the months of May and June, but earlier in the Southern States, the hibernating individuals leave their retreats, and those that have not mated the preceding autumn, now seek their mates, soon after which the females proceed to lay their eggs upon the crown of the food-plant, or beneath the ground upon the roots, where the early life of the insect is, for the most part, passed. At this season

*Dr. Fitch has observed pupæ in October which he did not doubt would pass the winter in that state (2d Report, p. 291).

of mating, and again in the autumn, large numbers are frequently seen upon the wing, and it is believed by Dr. Shimer and others, that at no other time do they indulge in flight, except at these two nuptial seasons, which, in Illinois, occur in May and August (see *Proceedings of the Academy of Natural Sciences of Philadelphia,* for May, 1867). This, however, is certainly an error, for there is reliable evidence of their migration upon the wing, for other reasons, apparently, than from want of food. See the statement of immense swarms flying eastward, in Iowa, on and about the 19th of June, in Prof. Forbes' 12th *Report on the Insects of Illinois,* p. 34.

From the gradual development of the eggs within the ovaries, and the long-continued oviposition, it follows that the final maturity of the insect is far from uniform. Hence it is that it is found in such different sizes and stages at any time throughout the season. The ripening of spring wheat may be given as about the time that the larger portion of the brood has acquired wings. During the wheat harvest they are often very abundant.

The second brood. — In July, eggs are laid for a second brood, which, according to observations made in Illinois, by Prof. Forbes, in 1882, were hatched about the middle of the month. Matured individuals were seen, the first, on August 8th; by the last of August, about one-half of the brood had matured, and the entire brood had become winged by the middle of September; by the 25th of September they had apparently ceased feeding. On the 3d of October they were rare in the corn-fields, and had evidently scattered in search of winter quarters.

As a general rule, this brood is less injurious than the first, as the corn and their other food-plants are more advanced and better able to resist attack.

Hibernation.—At the approach of cold weather they cease feeding, desert their food-plants and seek sheltered places where they may pass the winter. Some of them merely crawl down the stalks and stems and enter the ground, burying themselves therein to a moderate depth. Others take wing in search of secure retreats in any rubbish that may be found upon the borders or in the vicinity of the field, such as logs, old fences of wood or stone, coarse weeds, heaps of straw, leaves, etc. Stacks and piles of corn-stalks are particularly attractive to them, and are often employed as a lure, to serve as a holocaust later in the season for the myriads that have sought their shelter. Woodlands near the fields where they have abounded, furnish harborage for large numbers during the winter.

In the torpid condition in which they hibernate, they are capable of enduring an extreme degree of cold, and may even be frozen, it is said, without suffering harm.

A Dimorphic Form of the Chinch-Bug.

In addition to the ordinary form described and figured on page 150, there is also to be met with in limited numbers in some localities, a "dimorphic" form which has been characterized by Dr. Fitch as one of nine varieties observed by him, under the name of var. *apterus*. It is shown in figure 39. Prof. Riley has remarked of this and allied forms: "There are, as is well known to entomologists, many genera of the half-winged bugs, which in Europe occur in two distinct or 'dimorphous' forms, with no intermediate grades between the two, viz., a short-winged or sometimes a completely wingless type and a long-winged type. Frequently the two occur promiscuously together, and are found promiscuously copulating, so that they cannot possibly be distinct species. Sometimes the long-winged type occurs in particular seasons, and especially in very hot seasons. More rarely the short-winged type occurs in a different locality from the long-winged type, and usually in that case in a more northerly locality. We have a good illustration of this latter peculiarity in the case of the chinch-bug, for a dimorphous short-winged form occurs in Canada, and Dr. Fitch describes it from specimens received from the States as a variety under the name of *apterus* (*Second Report on the Insects of Missouri*, 1870, p. 22, f. 2; also, *Seventh Report on the Insects of Missouri*, 1875, p. 20, f. 4).

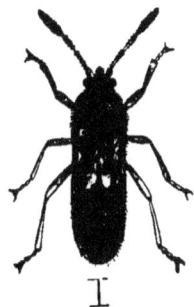

Fig. 39.— The Short-winged Chinch-bug.

Of a number of specimens sent to Mr. B. D. Walsh for identification, from Canada, in 1866, which had been taken from under the bark of an old log, all (eleven) were of this form, and were regarded by him as a "geographical variety" (*Practical Entomologist*, ii, p. 21)

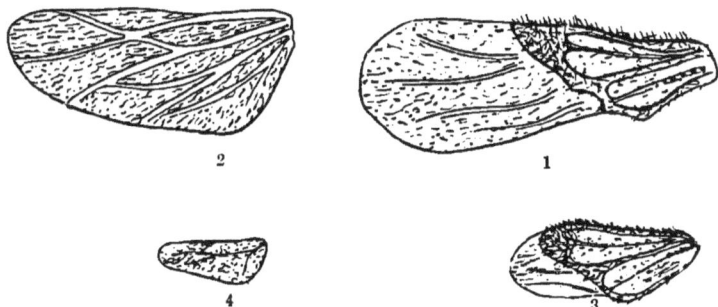

Fig. 40.— Wing-covers and wings of the chinch-bug, and its short-winged form, showing their nervulation: 1, wing-cover of normal form; 2, wing of the same; 3, wing-cover of short-winged form; 4, wing of the same. Enlarged to 18 diameters.

The same form was remarkably abundant among the specimens sent to me from Jefferson county, by Mr. Smith, and in those that were subsequently obtained by me upon the farm of Mr. King. At least twenty per cent of the entire number were of this form. The varietal name given it by Dr. Fitch is misleading, for none of them probably are wingless. Both the wing covers (hemelytra) and wings are very much shortened — the former usually to about one-half the normal length, and the latter to not much exceeding one-third. In figure 40, the relative size of the two is shown, together with the structure of each, from preparations and careful camera drawings made for this paper by Mr. C. E. Beecher. As the drawings are intended to represent structure only, the characteristic pigment spots and markings of the wing-covers are not given. Their coriaceous portion shows numerous short hairs, not extending beyond — the crinkled lines appearing elsewhere, and upon the wings, resulting from the unequal distension of the wing-membranes by the nerves. A comparison of the nervulation in the two forms cannot fail of being of interest to the student

Injuries of the Insect.

The chinch-bug is beyond all question one of the most destructive of our insect pests. It has been justly said of it: "The locusts of the west are the only creatures of this class which exist within the bounds of our national domain, whose multiplication causes more sweeping destruction than does that of this diminutive and seemingly insignificant insect."

Another writer, who has given it patient study, has remarked of it: "It is the most dangerous insect foe with which we have to deal. That it taxes them more heavily than all other such enemies combined, is burnt into the convictions of thousands of farmers by repeated heavy losses and bitter disappointment" (Forbes).

Throughout the Southern and Western States, or more properly, those lying within "the wheat-belt region," the chinch-bug is a well-known and dreaded enemy, from the almost incredible amount of injury which it inflicts, in certain years, upon the grain and corn crops. Probably the aggregate of pecuniary losses which have resulted to the United States from its ravages have considerably exceeded those inflicted by any other of our thousand insect pests.

That some idea may be had of the amount of these losses, a few of the more reliable estimates that have been made are herewith presented.

Mr. Walsh estimated the loss from the insect in Illinois alone, in the year 1864, at over $73,000,000, while a careful computation made by Dr. Shimer, for the same year, showed that in the extensive wheat and corn-fields of the valley of the Mississippi, three-fourths of the wheat and one-half of the corn was destroyed at a loss of more than $100,000,000, in the currency that then prevailed.

Dr. Le Baron, from data collected with great care, has estimated the value of the wheat, oats, barley and corn destroyed in the State of Illinois in 1871, at ten and a half millions of dollars; and for the six additional Western States of Iowa, Missouri, Indiana, Kansas, Nebraska and Wisconsin, approximately, double that of Illinois, making an aggregate in the North-western States of over $31,000,000 (*Second Report on the Insects of Illinois*, 1872, p. 144).

Prof. Riley, from county returns received from eighty-six counties in the State of Missouri, and from estimates of the remaining twenty-eight counties, has calculated the loss to the three staple crops of wheat, corn and oats alone, in 1874, in that State, at $19,000,000. He also states that during this year the losses from the same insect in Illinois may be safely put down at double that they were in 1871, or over $60,000,000 (*Seventh Report on the Insects of Missouri*, 1875, pp. 24, 25).

The years 1864, 1871 and 1874, from the warm and dry seasons that they offered, were unusually favorable to the multiplication of the insect, and were therefore years of its unusual abundance. Still it is not of rare occurrence that a single one of the wheat-growing Western States should suffer a loss of $10,000,000 in a single year. It seems almost incredible that the destructive powers of a small suctorial insect could be so great, but its enormous numbers must furnish the explanation. In times of its abundance it is so numerous as to cover the ground ; it blackens the stalks of the plants upon which it feeds; it fills the air when at seasons of its mating, it takes wing for flight; it marches to new feeding-grounds in solid bodies, upon and over one another; its invading armies sweep over and utterly destroy a wheat or corn-field in two or three days ; and the nauseous bed-bug odor which they exhale sickens those who are compelled to breathe it.

Only a little of our knowledge of this insect has been given in the preceding pages, but that space may be left for some notice of its recent operations in the State of New York, those who would know more of its interesting history, its habits, the natural enemies that prey upon it, and the various means used for destroying it, are referred to the excellent and extended papers upon it by Harris, Fitch, Thomas, Riley, Forbes and others which we have cited.

Operations in New York.

Its presence in New York in the years 1882 and 1883, while not attended with serious losses, was justly the occasion of great anxiety and fears. The attack, under its existing conditions, was of so threatening a character that I deemed it my duty to recommend most earnest efforts for its suppression. Just how far the efforts made in compliance with the recommendations aided in its arrest can only be surmised.

The following were my reasons for believing that the attack would continue and become more serious, as contributed to the *Albany Argus* of October 10, 1883:

The insect has planted itself, maintained a footing and has shown a rapid increase under unfavoring, unpropitious and unnatural conditions such as these :

First. It is regarded as a southern insect (extending farther northward, as do most animal forms, in the Mississippi valley), yet it has appeared in the most northern county of the State, and upon, if the report be reliable, the St. Lawrence river.

Second. Its attack has been made upon timothy. This seems to be its most unusual food-plant, and, therefore, we infer, the least suited to it. All previous accounts concur in giving it a preference for spring wheat above all things else; next in order, oats or corn, and last the grasses. Timothy is only mentioned as occasionally attacked by it.

Third. In all previous accounts, great prominence has been given to its being a hot and dry weather insect, dependent upon these conditions, not only for its multiplication, but for its existence. Heavy rains have been claimed to be invariably fatal to it. It could not abound, it is stated, in a wet season. Dr. Fitch had even made recommendation of sprinkling it with water (an artificial shower), as the best means for its extermination. In the present instance, the bug obstinately persists in multiplying, contrary to all rule. The past year and the present have both been years of excessive rain-fall in St. Lawrence county. Spring, summer and autumn have been exceptionally wet. In the spring, heavy and continued rains flooded meadows now showing the chinch-bug attack. At haying time, when the bugs were young, and according to all the statements hitherto made, readily killed by wet, the rains were so frequent and severe, that the grass cut could only be secured with difficulty. Upon Mr. King's farm, much of it was drawn in, upon favorable days, by improving the opportunity of extending the labor into hours after nightfall. At the present time grass is lying in fields in stacks, which could not be gathered, owing to continued rain; and fields of oats are still unharvested.

It appears that the insect has rapidly increased and largely extended its area during the present year, under conditions which should have been fatal to it. Why it has been otherwise may perhaps find its explanation in the fact that it is a new introduction into this part of the United States, and that it is following the law well known to prevail in the introduction from abroad (Europe principally), of nearly all of our injurious insects. With scarcely an exception, with their importation, they become far more destructive, causing greater ravages and often attacking new food-plants.

As the past history of the insect has shown that parasites and other enemies have entirely failed to arrest its multiplication, we are compelled to believe, from present indications, that it has come to stay, and that it will do so, unless effectual means are taken to prevent it. Its capability of increase is wonderful. Under the most conservative circumstances, a single chinch-bug, depositing its eggs about the first of June, would be, in the following August, the progenitor of a quarter of a million.

It should not be necessary to urge the importance of doing whatever can be done to arrest this attack, which threatens to be more serious to New York than was that of the wheat-midge, the loss from which, in some years, was computed at $15,000,000.   If it should continue to increase it will doubtless extend to wheat, and corn and other of the grains. In its southern extension in this State it would naturally become more serious.   At the present it is known in but two counties — Jefferson and St. Lawrence.   It seems practicable by prompt, earnest and combined effort. to prevent its extension and to check it where it now exists.

Observations upon the Attack.

In addition to the information contained in the communication of Mr. Smith, given on page 149, I am able to add the following, from my personal observations, made during a visit to Hammond, and on inspection of the infested farm of Mr. King, on October 5th and 6th.   The cold weather of the preceding few days (ice was formed upon three nights) had doubtless driven most of the bugs to their winter quarters for hibernation in crevices, beneath boards, rails, etc., in rubbish heaps, and to many other secure retreats where such insects are accustomed to hide. Yet upon parting the roots of the timothy upon borders of the killed portion in one of the fields they were found in alarming numbers — in some spots sufficient to cover the ground with their bodies over an area of two or three inches in diameter, having perhaps congregated for warmth in such places.   In one spot, upon the warm slope of a dead furrow, they could be seen, in large numbers, running like ants over the ground. Elsewhere, they were concealed among the roots, near to and about the bulbs, upon which they appeared mainly to feed.   Their presence in any spot could always be detected by bringing the nose near the ground by the peculiar bed-bug odor that they exhaled.   This method of detection proved more convenient and infallible than looking for them.

The territory occupied by them was more extended than was at first supposed.   Nearly all of the farms in the neighborhood of Mr. King had been attacked, some of them the preceding year, and discoveries of attack not before suspected were, upon examination, being made daily. A range, at the time, of about eight miles was indicated.   It was believed to occur throughout most of the town of Hammond, and to extend into Alexandria upon the St. Lawrence river.

The following are a few of the memoranda (the remainder lost) of the number of acres showing the attack in the town of Hammond, upon farms in the immediate vicinity of Mr. King, which were made at the time of my visit:

Mr. H. C. King, sixteen acres of timothy and clover; cut on the 5th of July; the attack first noticed in August; more than one-half of the timothy was eaten. On another portion (upland) of his farm, three acres of timothy, and in another, twelve acres of meadow were eaten.

Mr. E. J. White, first noticed the injury in September of 1882, in the killing of the timothy upon a spot between two and three rods square. In 1883, it was again observed after haying in July, and continued to extend, and to appear in three other places — in all, seven or eight acres.

Mr. Keller Dygert, adjoining Mr. King's farm, over six acres destroyed.

Mr. Robert Rogers, five or six acres of timothy very badly injured.

Mr. Robert Broidee, a meadow badly infested, but extent not stated.

Mr. Abel Pickett, a small spot, commencing the previous year, and is steadily extending.

Mr. Wm. Cuthbert's farm, five miles from Mr. King, near the St. Lawrence river, reported as showing the attack of the insect.

Mr. Thomas B. Phillips had two acres injured the preceding autumn; present area considerably larger but not given.

Mr. Charles Fitch, of the town of Morristown, six acres destroyed.

A wheat-field of Mr. King seems to have been infested, but to have been checked by its roots having been submerged by a heavy rain-fall continuing for several days.

The insect is reported on the farm of Mr. Cook, in the town of Champion, Jefferson county, in the vicinity of Pleasant lake, and near the northern line of Lewis county.

Measures Recommended to Arrest the Attack.

In view of the severity of the attack, and the seeming need of earnest efforts to arrest it, it was thought proper that general attention should be called to it throughout the infested region, and instructions given as to the best means for arresting it, particularly to such as could be at once resorted to, in order to reach as large a portion as possible of the autumnal brood. A circular giving such information was accordingly prepared, and an edition of three thousand copies printed and distributed throughout the portions of the State where the attack was observed, together with such contiguous territory as it might be expected to reach in another season.

As a portion of the history of the attack, and as containing directions which would be of service in future appearances of the insect in New York or in the New England States, the circular is herewith given

Circular No. 1 — October, 1883.

NEW YORK STATE MUSEUM OF NATURAL HISTORY, ⎱
DEPARTMENT OF ENTOMOLOGY. ⎰

Directions for Arresting the Chinch-bug Invasion of Northern New York.

Portions of St. Lawrence county, New York, are now suffering from a serious attack of the chinch-bug (*Blissus leucopterus*) — perhaps the most injurious of our insect enemies.

It has already, in the third year (probably) of its introduction, and the second year of the observation of its attack, spread to such an extent, and shown such a rapid increase under very unfavorable conditions, that a continued increase in its diffusion and destructiveness is probable, unless effectual measures can be taken to prevent it.

At present, only timothy and other grasses seem to have been attacked. Wherever attacked, the root is destroyed, and the grass, consequently, is entirely killed.

With its increase, its ravages would extend to wheat, rye, barley and corn, which are its favorite food-plants.

Its extension over the State of New York, as now threatened, would be attended with an annual loss of millions of dollars.

It seems practicable, at this stage, to prevent this extension, by earnest and combined effort throughout the district now invaded.

The most favorable time for this effort has already passed ; but much may be accomplished by immediate action.

As it is of very great importance that this destructive insect — the terror of our Southern and Western farmers — should not be permitted to obtain a permanent footing in our State, hitherto free from its depredations, a prompt and full compliance with the following directions is strongly urged :

1. Let every farmer in St. Lawrence county and adjacent counties in Northern New York (particularly in the western portion of St. Lawrence and northern of Jefferson), examine his meadows for patches of dead grass, looking as if winter-killed, indicating the attack of the insect. As an aid to its ready recognition, the infested areas upon the farm of Mr. H. C. King, of the town of Hammond, St. Lawrence county, may be examined.

2. If the attack is detected, burn the dead grass and its surrounding border of fifteen or twenty feet not yet showing attack. This may be effectually done by first applying a covering of straw. A favoring wind is desirable for the purpose.

3. Plow the burned area (better still if the plowing extends beyond this limit and embraces the entire meadow) in broad and deep furrows, turning the sod completely and flatly over, not permitting it to lie in ridges.

4. To insure the more effectual burying of the insects that may be at present feeding upon, or preparing to pass the winter among, the roots of the grasses, harrow the plowed surface slightly, and follow with a heavy rolling.

21

5. Where the meadows will not permit of plowing as above, gas-lime, wherever it can be conveniently obtained from the gas-works at Ogdensburg, Watertown, etc., may be distributed over the ground, at the rate of two hundred bushels to the acre. The gas-lime would also serve as a valuable fertilizer.

Of the above directions, the first four should be followed *at once*. The application of gas-lime might be postponed until the month of November, before the setting in of winter, or to the early spring. It should be confined to the dead and infested portions of the meadows, as in its fresh state it would kill the grass. In the winter, during February, it may safely be distributed over the entire fields, where it would probably serve the additional purpose of a preventive of a spring attack.

New attacks and more widespread distribution may be looked for about the first of June in the ensuing year. Directions for meeting these, by other methods, will be given hereafter.

It is hoped that every one interested will cheerfully comply with the above directions, and not render necessary a resort to compulsory legislation, which would undoubtedly call for a large increase of labor and expenditure. The agricultural interests of the State of New York may justly demand that, if possible to prevent it, the chinch-bug shall not be allowed to gain a permanent footing as a grain and grass destroyer within its borders. Its injuries in the State of Illinois, in a single year, were estimated at $73,000,000 — almost five times the amount computed for the wheat-midge ravages in New York, at the time of its greatest destructiveness.

OFFICE OF THE STATE ENTOMOLOGIST, *October* 18, 1883.

To the above circular the following figure (after Fitch), and description was appended, to aid in the recognition of the insect.

The CHINCH-BUG in natural size and as enlarged (about ten diameters). Color: black, with white wing-covers, having a black subtriangular spot on the outer margin of each, and two black veins nearer the base. The legs, the sucking-tube, and the base of the antennæ, are deep honey-yellow; the feet and the last joints of the antennæ are black. Length, about three-twentieths of an inch.

The young, appearing early in June and late in August, are blood-red, with a white band across their middle: later they change to brown and afterward to black.

Remedial Measures Employed.

In a report made by me to the Regents of the University of the State of New York, under date of January 8, 1884, and published in the Thirty-seventh Annual Report on the State Museum of Natural History, in referring to the above circular I stated as follows :

I regret to have to report that the response given to the directions of the circular have fallen short of their requirement. Plowing under the infested areas has been quite general, but I do not learn that it has approached the thorough character recommended. Burning has not been resorted to, except upon the farm of Mr. King. The application of gas-lime will probably not be made, to any great extent, as it is reported as not easily to be obtained. Perhaps no other result should have been anticipated at this stage of the attack, or before the absolute necessity of vigorous action should be unmistakably apparent. Former experiences show that our farmers, as a rule, are indisposed to yield ready compliance with recommendations simply, although calculated to save them from serious pecuniary loss, particularly if such recommendations involve any expenditure beyond that of quite a limited amount of extra labor on their part. It would, therefore, seem to be a wise economy for the State, whenever a continued extension of any formidable insect attack presents itself, that a prompt resort be had to effectual preventive measures, through legislation compelling the action desired and not otherwise to be had. Several laws for the prevention and destruction of injurious insects exist upon the statute books of European countries. In our own State and others there are laws against noxious weeds; and it would indicate enlightened progress if there were also those controlling the unlimited spread of some of our more harmful insect pests.

Cessation of the Attack.

Although the insect was reported by Mr. King to have been found by him, in alarming numbers, late in October, hidden beneath rails, chips, bark, etc., yet the apprehension of its increase and spread in the spring of 1884 was not realized. Doubtless the general plowing of the infested lands aided materially in its destruction; yet to the more effective operation of natural causes, such as the extreme cold of the winter or the early spring rains, or other unfavorable meteorological conditions, must be credited, mainly, the arrest of the attack, either through the death of the mature insects during the winter, or that of their progeny in the early spring.

So far as I could learn, the insect did not reappear to the extent of committing serious injury. In localities where it had abounded the preceding year, and the land had not been plowed, its presence, in hibernated individuals, was observed as soon as the snow was gone. No further damage was reported to me, except in one instance where no attention had been paid to the recommendation of thorough autumn plowing. Here a piece of wheat of several acres in extent was attacked, and considerably injured by it.

Observed in other Northern and Eastern Localities.

During the spring of 1883, the insect also occurred abundantly in at least one locality in Massachusetts, for according to Dr. George Dimmock, "on the 28th of March, the lowland between Belmont and Cambridge was swarming with them" (*Psyche*, Novem.-Decem., 1883, iv, p. 119).

An earlier observation of it in Massachusetts has been communicated to me by Dr. Packard, in the following extract from his diary: "June 17, 1871, at Salem, Mass., chinch-bugs with wing-covers extending over the basal third of the abdomen, seen in copula, end-to-end." Dr. Packard has also published the fact of his having taken the insect "frequently in Maine and even on the extreme summit of Mt. Washington in August" (*Guide to the Study of Insects*, 1869, pp. 543–4).

Mr. Henry L. Fernald, of Orono, Maine, who is making special study of the Hemiptera, has given me the following memoranda of its occurrence at Orono, latitude 44° 40′: "One example captured in 1879; one on May 25, 1880; three on June 3, 1882. Since that time quite common in June; taken in all cases by sweeping over grass."

Its occurrence in Canada, in 1866, in a few examples has already been referred to (page 155). Upon inquiry, I learn that the locality of their collection by Mr. J. Pettit, was at Grimsby.

For a still more northern locality of the insect we are indebted to the recent collections of Mr. W. H. Harrington, of Ottawa, Ont., who, at the annual meeting of the Entomological Society of Ontario, at London, in October of 1884, exhibited specimens of the chinch-bug which he had found abundant at Sydney, Cape Breton (N. latitude 46° 18′) during a visit there in September (*Canadian Entomologist* for November, 1884, xvi, p. 218).

Largus succinctus (Linn.).

The Margined Largus.

(Ord. HEMIPTERA: Subord. HETEROPTERA: Fam. PYRRHOCORIDÆ.)

Cimex succinctus LINN.: Syst. Nat., ii, 1767, p. 727, No. 82.

Cimex rubrocinctus DE GEER: Mem. Hist. Ins., iii. p. 339, No. 13, pl. 34, f. 19.

Cimex succinctus FABR.: Spec. Ins., ii, 1781, p. 369, No. 185; Syst. Ent., 1775, p. 723, No. 133; Maut. Ins., ii, 1787. p. 303.

Lygæus succinctus FABR.: Ent. Syst., iv, 1793, p. 170.

Largus succinctus HERR.-SCH.: Wanz. Ins., vi, p. 78, f. 648.

Capsus succinctus var. A. SAY: Heterop. Hemip N. A., 1831, p. —; in Trans. N. Y. St. Agricul. Soc., xvii, 1858, p. 783.

Largus succinctus. GLOVER: in Rept. Comm. Agr. for 1875, 1876, p. 124, f. 28; MS.
 Notes Journ.—Hemipt., 1876, p. 43, pl. 1, f. 12; MS. Notes Journ.—Cotton
 Insects, 1878, pl. 16, f. 9 (note to figure).
Largus succinctus. UHLER: in Bull U. S. G.-G. Surv. Terr., i, 1876, p. 315; List
 Hemiptera W. Miss. Riv., 1876, p. 49; Stand. Nat. Hist., v, p. 288, f. 331.
Largus succinctus. LINTNER: in Count. Gent., xlvi, 1881, p. 663.

The injurious habits of this species have only, up to the present,
been noticed in Texas. A letter received from San Antonio, represents
it as quite annoying in a peach orchard, from its working into the fruit
and spoiling it just before its time of ripening.

Distribution and Variations.

Although not as yet recorded from the State of New York, it without
much doubt occurs within its limits, as it is found in Pennsylvania
and New Jersey, and has a broad distribution over the United States.
Mr. Uhler gives as its habitat, " Pennsylvania to Florida, and westward to Texas, Arizona and southern Colorado. The western specimens are blacker and not so brightly red-margined as those from the
coasts of Georgia and Florida. In the sea islands of the latter a variety
occurs which is of a dirty sand-red.

" The genus is essentially American, and ranges between the northern warm-temperate zone and the southern warm-temperate zone. The
insular and equatorial ones of the lowlands are marked with yellow
spots, while the others are more uniform and plainer in their pattern."

Its Appearance.

It bears a marked resemblance to the common squash-bug, *Anasa
tristis* (De Geer), Fig. 42, in size and form; the thorax is very nearly the
same in shape. It measures one-half of an inch in length by one-fifth in
breadth, is of a rusty black color, with the thorax and upper wings

FIG. 41.— The
Margined Largus,
LARGUS SUCCINC-
TUS — enlarged
one-half diameter

freckled and broadly bordered with red. The brief diagnosis of Fabricius gives its characteristic features: "Oblong,
the thorax, margin of the elytra, and base of the femora
red." The diagnosis of De
Geer adds to the above, "ashy-
black, femora toothed in front."
The insect is represented in
Fig. 41. The variety described
by Say, as inhabiting Mexico,
has the " surface paler, with numerous black
punctures, giving a dusky appearance; origin of the antennæ and a line on each side
of the origin of the rostrum sanguineous."
Uhler has referred this variety to *Largus*
cinctus Her.-Schf., which as it differs only in a slight degree from

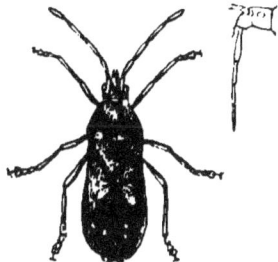

FIG. 42—The Squash-bug, ANASA TRIS-
TIS — enlarged one-half diameter; head
and beak still more enlarged.

L. succinctus, "will no doubt hereafter prove to be only the western form of it." This would extend the range of the species into California, Oregon, Nevada and Arizona.

In its early stages, according to Uhler, it is of a brilliant steel-blue color, with reddish legs, and a bright red spot at the base of the abdomen.

Habits.

Scarcely any thing is known of the habits of this insect, as it has not engaged the attention of our economic entomologists. Mr. Glover had found it hibernating in Maryland, under moss, stems and bark in mid-winter but had never seen it injuring plants. In the same State and in Virginia, the adults have been observed along the borders of oak-woods in the months of July and August. It also occurred in the southern cotton-fields, occasionally upon the bolls, but commonly on the ground and under stones. It had been represented as stinging severely with its proboscis and as destroying other insects.

It is closely allied to the well-known red-bug or cotton-stainer of the Southern States, *Dysdercus suturellus* Her.-Schf.,* which is so injurious to cotton, in sucking the sap from the plants and the bolls, and staining the fibres of the opened bolls indelibly with their excrements, to the extent of greatly impairing the market value of the cotton.

Other members of the family, although less nearly allied to it than the above, and therefore of but little value in indicating habits, are the chinch-bug, *Blissus leucopterus* (Say), the false chinch-bug, *Nysius angustata* Uhler (= N. destructor *Riley*), and *Lygæus turcicus* Fabr.— the latter a common New York species, found abundantly upon milk-weed (*Asclepias*), during the month of August, and said to have been seen feeding upon the caterpillars (? *Euchetes egle*) infesting the plant.

The habits of a European species *Pyrrhocoris calmariensis* Fallen, from its near relationship, are probably much like those of *L. succinctus.* Prof. Westwood had found them swarming in gardens in the neighbor-hood of Berlin, where they were engaged in sucking fallen berries and seeds, as well as such of their companions as had been trodden under foot (*Introduc. Classif. Ins.,* ii, p. 481, fig. 121, 8, 9). Haus-mann had observed their partiality for dead insects, and that they would not attack living ones.

---

* See Glover's "MS. Notes from my Journal—Cotton and the Principal Insects Frequent-ing or Injuring the Plant," 1878, where the transformations, structure and habits of the insect are illustrated in the fourteen figures of Plate XVI, and mention is made of the cotton plants near Jacksonville, Florida, in 1855, being literally colored red with the multitudes that were crawling over the stalks, leaves, and bolls.

Remedies.

There is but little probability that this insect will become particularly injurious to peaches or to other fruits. From the little that is known of its habits, and from what may be inferred from related species, it will rarely, if ever, ascend trees in search of food, but will prefer to seek its food upon the ground.

It would be difficult or impossible to devise a preventive of its attack, as its suctorial habits preclude the beneficial use of external applications to the fruit. It remains, therefore, in the event of its abundance and threatened injuries, to find means for destroying such numbers as to lessen the amount of its depredations. It is probable that the same methods that have been employed with beneficial results for the destruction of the cotton-stainer, as given by Mr. Glover, will also serve for this: "These insects being in the habit of collecting together where there were splinters or fragments of sugar-cane on the ground, advantage was taken of this fact to draw them together by means of small chips of sugar-cane laid upon the earth near the plants, when they were at once destroyed by boiling water. They also collect around heaps of cotton seed where they may readily be destroyed at the commencement of cold weather. Small heaps of refuse trash, dried cornstalks, or especially of crushed sugar-cane, may be made in various parts of the plantation in the vicinity of the plants. Under these the insects take shelter from the cold, and when a sufficient quantity of the bugs are thus drawn together the various heaps may be fired, and the insects destroyed with the trash. A very cold morning, however, should be selected, and the fire made before the insects have been thawed into life and vigor by the heat of the sun; and especially all dead trees, decayed stumps, and weeds in the vicinity of the field should be burned or otherwise destroyed, as they afford a comfortable shelter for all sorts of noxious insects, in which they can pass the winter in a semi-dormant state."

Cicada septendecim Linn.

The Seventeen-year Locust.

(Ord. HEMIPTERA: Subord. HOMOPTERA: Fam. CICADIDÆ.)

LINNÆUS: Syst. Nat., i, pt. ii, 1767, p. 708, No. 20.
FABR.: Syst. Ent., 1775, p. 679, No. 6; Spec. Ins., ii, 1781, p. 319, No. 6; Mant. Ins., ii, 1787, p. 266, No. 9 (*Tettigonia septendecim*).
HILDRETH: in Silliman's Amer. Journ. Sci.-Arts, x, 1826, pp. 327–329 (emergence, oviposition, injuries, etc.).

Morris: in Proc. Acad. Nat. Sci. Phil., iii, Nov., 1846, pp. 132–134; Ib., March, 1847, pp. 190–191 (larvæ on roots of fruit trees); in Downing's Horticulturist, ii, 1847, p. 16.

Burnett: in Proc. Bost. Soc. Nat. Hist., iv, 1851, p. 71 (sexual system and musical apparatus); Ib., p. 111 (appearance in cleared lands).

Harris: Ins. N. Eng., 1852, pp. 180–189; Ins. Inj. Veg., 1860, pp. 206–217, f. 87, pl. 3, f. 7.

Fitch: in Trans. N. Y. St. Agricul. Soc. for 1854, xiv, 1855, pp. 742–753; 1st Rept. Ins. N. York, 1856, pp. 38–49.

Walsh: in Pract. Entomol., i, 1865, p. 19 (locust districts, from Fitch).

Riley: 1st Rept. Ins. Mo., 1869, pp. 18–42, figs. 6–13; 4th Rept. do., 1872, pp. 30–34.

Le Baron: 2d Rept. Ins. Ill., 1872, pp. 124–133.

Packard: 3d Ann. Rept. Ins. Mass., 1873, pp. 16–20, figs. 142, 143.

Bessey: in Amer. Entomol., iii, 1880, pp. 27–30 (distrib. in Iowa).

Saunders: Ins. Inj. Fruits, 1883, pp. 35–39, figs. 24–27.

The Cicada in Western New York.

Examples of the above-named insect, together with peach-twigs showing its work, and empty pupal cases attached, were received at this department, about the last of June, 1882, from Ontario county, New York, with the inquiry if they were the true seventeen-year locust, and if so, if their appearance only at such long intervals could be accepted as a scientific fact.

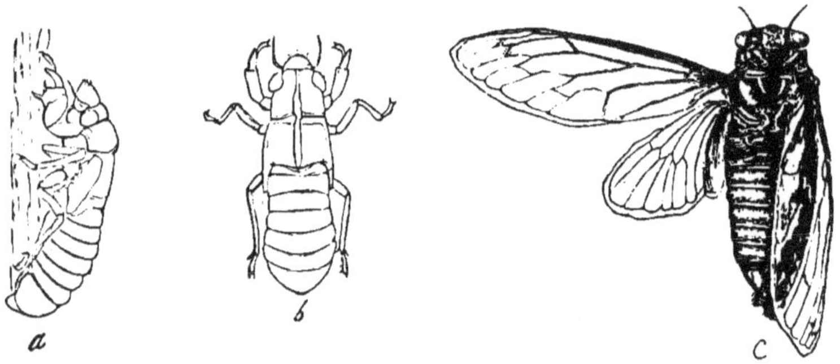

Fig. 43—The seventeen-year Cicada — Cicada septendecim: a, the pupa; b, the pupa case; c, the mature insect.

Fig. 43 shows the insect with its wings spread upon one side, the pupa from which it is disclosed, and the pupal case.

Seventeen Years Required for its Transformations.

Notwithstanding all that has been written of this species, and the labor that has been expended upon its study, it is a matter of surprise that there should still prevail a wide-spread incredulity that the period

stated is really required for the development of the insect from the egg
into the perfect winged state. Yet it is a fact well established in
science, and must therefore be admitted, although among the three
hundred and fifty thousand species of insects known to us, no other has
been found to require so long a period for its changes.

The Insect Seen Almost Annually.

It is sometimes urged by those who are unable to appreciate the au-
thoritativeness of a scientific statement, that we hear of the appearance
of these locust visitations at much shorter intervals — in some parts of
the United States almost every year.

This is true. To go no further back in our records than to the year
1850 — from that time to the present the periodial Cicada (it is improp-
erly called a *locust*) has made its appearance in portions of the United
States, in each subsequent year to the present included, with the
single exception of the year 1873. It will continue to appear during
each subsequent year of the century, with three exceptions — 1887,
1890 and 1892, which in the chronology of visitations, happen not to
be "locust years."

Why Reported so Frequently.

The explanation of these almost annual appearances of the species
is very simple, and as follows : There are a number of distinct "broods"
within the United States, each embraced within certain geographical
localities and limits (although often overlapping one another) in which
they appear, and within which they are always true to their appointed
time.

As the limits of these broods do not vary to any considerable extent
in their successive returns, it follows that they do not cover the entire
country, but that there are portions in which they are never seen.

A Thirteen-year Brood.

Not all of these broods, however, are confined to a seventeen-year
cycle. The researches of Professor Riley have shown that in the Southern States, and extending as far northward as southern Illinois, there
are also broods that occur regularly at intervals of *thirteen years*. to
which he has given the name of *Cicada tredecim*. No specific difference, however, can be discovered in the features of the two broods,
and he has, therefore, expressed his opinion that the latter is not entitled to be regarded as a distinct species, but that it should rather be
considered as a race, or as an incipient species, to which for convenience it is desirable to give a distinctive name (*Supplement to the
Missouri Reports*, 1881, p. 58).

22

Number of Broods in the United States.

Dr. Fitch, in his *First Annual Report on the Insects of New York,* recorded nine broods in different parts of the United States. Prof. Riley, as the result of a more thorough and extended search of old records, gives, in his *First Report on the Insects of Missouri,* twenty-two broods, one of which he has subsequently rejected, from its failure to appear at its expected time and place, and the doubtful evidence upon which its former appearance (in 1870) rested. Of these twenty-one broods, seven are of the thirteen-year race.

Broods in the State of New York.

Of the above, five occur within the State of New York, viz.:

1. Commencing with the one, the appearance of which in Ontario county in June of 1882, as has been stated; former appearances are recorded in 1797, 1814, 1831, 1848 and 1865. Its next return will be in 1899. Its probable distribution is in portions of Wyoming, Monroe, Livingston, Ontario, Yates, Seneca, Cayuga and Onondaga counties.

2. To appear in 1885, in Brooklyn and the western part of Long Island, in the vicinity of Rochester, N. Y., and probably to a lesser extent in intermediate portions of the State.*

The range of this brood is stated to be from South-Eastern Massachusetts (across Rhode Island and Connecticut) to Long Island, New Jersey, Pennsylvania, extending up into Western New York, southward into Maryland, Virginia, West Virginia, Ohio, Michigan, Indiana and Kentucky. Its former records are, 1715, 1732, 1749, 1766, 1783, 1800, 1817, 1834, 1851 and 1868. The next appearance will be in 1902.

3. A brood to appear in 1889, on Long Island. Former visits have been in 1838, 1855 and 1872. It will also return in the year 1906.

This is of very extensive range, commencing in Massachusetts and crossing Long Island, is continued in New Jersey, South-Eastern Pennsylvania, Delaware, Maryland, Virginia, West Virginia, down the Ohio river and " the valley of the Mississippi river, probably to its mouth, and up its tributaries into the Indian Territory."

4. A brood in 1894, in the valley of the Hudson river. Dr. Fitch gives its northern extension as in the vicinity of Fort Miller, near to where the river has its westward trend. It extends southward along the river to its mouth, embracing, in its lateral extensions, New Haven, in Connecticut, Northern New Jersey and North-Eastern Pennsylvania.

---

*In the *New York Sunday Press* of June 20, 1885, the following notice appears: " Locusts have made their appearance by millions, in the middle of Long Island and on Staten Island. In Prospect Park, Brooklyn, the nuisance is said to be intolerable. In many places the insects cover the roadway and are crushed under wheels in countless numbers."

What is apparently the same brood seems also to extend into Maryland, Virginia and North Carolina; also into Michigan and Indiana, or perhaps, these last may be a coincident brood.

The records of this, date back to over a century and a half, viz.: to 1724, in New Haven, Conn. Subsequent returns were in 1741, 1758, 1775, 1792, 1809, 1826, 1843, 1860 and 1877. In the last recorded year they were very abundant in the vicinity of Albany. Large numbers of these were found to be infested with a fungus, which was described and named by the State Botanist, Prof. Charles H. Peck, as *Massospora cicadina*.

5. A brood in 1900 in Western New York. Former years of its appearance were 1832, 1849, 1866 and 1883. Its range is apparently not very extensive, as it seems to be limited to Western New York, Western Pennsylvania and Eastern Ohio. It is Dr. Fitch's 2d brood; Riley's No. xx. Miss M. H. Morris, of Germantown, Pa., to whom we are indebted for many valuable observations upon this insect, has recorded it as occurring in the year 1849 "in the northern portion of New York, from Buffalo through the entire length of Genesee county" (*Proc. Bost. Soc. Nat. Hist.*, iv, 1851, p. 110). This would give it a distribution at that time centrally through Erie, Wyoming and Genesee counties, with probable occurrence in Niagara, Orleans and Monroe counties. Since that time its range seems to have become more restricted and its numbers reduced. Our only knowledge of its appearance in 1883 is from an item in the *New York Herald* of about July 10th, that a "swarm of locusts," doing much damage, had appeared in Chautauqua county, the extreme south-western corner of the State.

In a former publication, I included with the above, a sixth New York brood — the xviiith of Prof. Riley — of which he predicted the return in 1881, of a few in Westchester county, based upon the observation by Mr. James Angus, of "straggling specimens" in Westchester county in 1864. But, as Mr. Angus informs me that in that year he "saw a few odd ones only, not more than two or three, although hunting pretty closely," and that in 1881, none were seen by him, it would obviously be improper to cite the above very limited appearance as a New York brood.*

In the above statement of years of appearance and range of distribu-

---

[*Its return was, however, noticed by Mr. W. T. Davis, upon Staten Island, N. Y., who reports, in *Entomologica Americana*, i, 1885, p. 92: "On May 8, 1881, while collecting insects with Mr. Leng in the neighborhood of Watchogue, Stat. Isl., we found a red-eyed Cicada pupa under a stone, and on June 5, eight specimens were collected, all of them males, and many of them being wet, having recently emerged. By the 12th of June they had become quite numerous, and I noted at the time that about one tree, I counted fifty-two pupa skins of the red-eyed Cicada."]

tion, we have drawn almost entirely upon the valuable compilation of the Cicada broods occurring in the United States given by Prof. Riley in his First Missouri Report, a work based upon so careful and extensive research among published records as to have left room for but little subsequent correction or addition.

The 1865–1882 Brood in New York.

There was not the opportunity of ascertaining from personal observation the range of this brood and the extent of its ravages in the summer of 1882, in Central and Western New York, and, subsequently, efforts to obtain information upon these points have met with hardly any success. The counties in which it probably occurs are given upon page 170. Its greatest abundance seems to have been in the vicinity of Canandaigua lake.   In the *Ontario County Times* of June 28th, we find the following:

The fruit growers of Vine Valley [Yates county], on the east side of Canandaigua lake, are seriously disturbed by the recent appearance in that locality of countless millions of "seventeen-year locusts." They have come in such immense numbers as to cause widespread anxiety. Our informant, who visited the infested district on Monday, reports that the peach orchards and vineyards are suffering much from the ravages of the locusts, and says the noise made by their movements is so great that the human voice is almost drowned by it.   They are feeding upon the tender foliage of the trees and vines, and fears are entertained that the growing fruit will be destroyed.   The locusts have also appeared on the west shore of the lake [in Ontario county], and are reported to be doing considerable damage in peach and apple orchards.

Mr. N. J. Milliken, editor of the *Ontario County Times*, communicates the information that the locusts were particularly destructive in the Middlesex valley, on the east shore of Canandaigua lake, Yates county.   They occurred in the Lake Keuka region, but no great amount of harm was done by them.   In the town of Terry, of the same county the injury committed was comparatively slight, although " the noise made by them was equal to the united drum-corps of a mighty army."

In the last-named town, according to a note communicated by Prof. Riley, the area occupied was about four square miles, and somewhat less in the town of Middlesex.

Mr. Thomas J. Powell, of Naples, N. Y., has informed me that on the 28th of June, of 1882, he was in Victor, Ontario county, and went north four or five miles and found the locusts very abundant in the trees, making a continual humming.   So far as he could learn they had caused but little damage.

Time of Appearance.

The time of the appearance of the insect would, of course, vary with the latitude, the southern ones being several weeks earlier than the extreme northern. In general terms, it may be stated as in the latter part of May. Dr. Harris states that "in Alabama, they leave the ground in February and March; in Maryland and Pennsylvania in May, but in Massachusetts they do not come forth until the middle of June."

In the extensive brood of 1868, which reached from the Atlantic to west of the Mississippi river, individuals were first noticed in Maryland and Pennsylvania from the 20th to the 25th of May; in Illinois on the 19th of May, and at St. Louis, Missouri, on the 22d of the same month. They usually continue for from five to six weeks, although straggling specimens may sometimes be seen for weeks later.

Life-History.

For detailed description of the egg, larva, pupa and perfect insect the First Report of Dr. Fitch may be consulted, and for the natural history, the excellent volume of Dr. Harris. It may suffice for this pres-

Fig. 44. — Cicada septendecim, showing a at,the beak; at b, the ovipositor.

ent notice, to state that the female selects a small branch of an apple-tree, oak or some other hard wood, near its tip, into which to place her eggs, always taking a position with her head directed toward the trunk. With her ovipositor shown at b, in Fig. 44, she saws little slits in the twig, making an oblique hole to the pith, and enlarging it into a fissure with splintered outward edges, as represented in Fig. 45. In each of these she places from ten to twenty eggs, in pairs side by side, but separated from each other by portions of woody fibre, and inserted somewhat obliquely so that their ends point upward. About fifteen minutes are required to make a fissure and fill it with eggs. Moving backward toward the tip of the branch another and others are made in line — sometimes to the number of fifty, if the twig be favorable for the purpose, by the same insect. Her complement of eggs is from four to five hundred. The female, exhausted by her labor, soon thereafter drops from the tree and dies.

The time required for the hatching of the eggs has been variously stated at fifty-two days, forty-two days, and even so low as fourteen days.

Fig. 45. — Twig punctured by the Cicada.

The newly-hatched Cicadas — about the one-sixteenth of an inch long, and shown in Fig. 46 — are slender, grub-like creatures, lively as ants, but after running about upon the tree for a short time, they

Fig. 46. — Newly-hatched larva of Cicada septendecim.

gather up their limbs and drop to the ground. Here they at once bury themselves, by the aid of their strong fore feet, which are admirably fitted for the purpose, and work their way, until they can attach themselves to the tender and succulent rootlets, into which they insert their beak, and find abundant nourishment in the sap. Ordinarily they bury at a moderate depth below the surface, where they remain until their advanced growth compels them, when in excessive numbers, to descend deeper in search of an ample food-supply.

During the years of larval growth, but little alteration takes place in their appearance, beyond an increase of size, except the gradual development and enlargement of the four scales upon the back and sides that are to contain the wings in the pupal stage.

When near maturity, the larvæ gradually ascend toward the surface through circuitous galleries of about five-eighths of an inch in diameter, the walls of which are firmly compacted, and according to a statement, cemented and varnished so as to be water-proof. The upper portion alone of these burrows, to the extent of six or eight inches, is empty. The

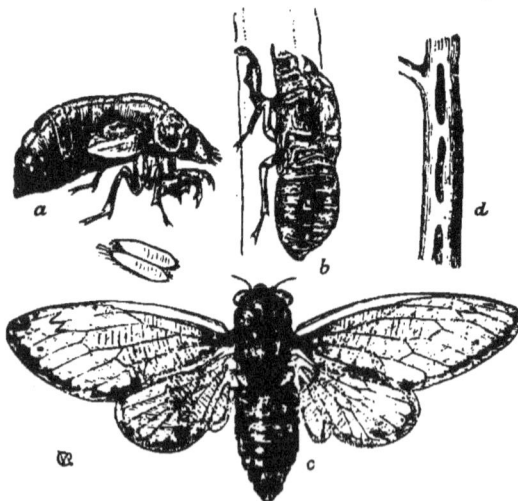

Fig. 47. — Cicada septendecim: a, the pupa; b, empty pupal case; c, the perfect insect; e, the eggs.

larva remains herein during its pupal state, and until the time for its final transformation. When this has fully come, the pupa, shown at a, in Fig. 47, burrows upward through the ground during the night, ascends the branch of a tree or some other convenient object, to which it fastens itself securely by its claws. Remaining in this position for awhile, and until the full time for

its change is at hand, the pupal case parts in front and upon the back, the Cicada emerges through the fissure, crawls upward a little distance, leaving the empty case tightly fastened to the trunk as at b, the wings expand, all the organs have become fully developed, and the insect has attained its perfect stage, as shown at *c*. If food is needed, it punctures the twigs with its beak, shown at *a*, in Fig. 44, and feeds upon the sap. After two or three weeks of a merry life, if we may judge from the almost ceaseless music with which the days are made vocal, the important work of oviposition is commenced and soon completed. Another month, and the brood, of so many years' gradual development, has become extinct.

Its Music and Musical Apparatus.

Dr. Fitch has referred to "the discordant din of the Cicada's shrill song." In endeavoring to convey an idea of the peculiar note, he says — "it is repeated at short intervals, and may be represented by the letters *tsh-e-e-E-E-E-e-e-ou*, uttered continuously, and prolonged to a quarter or a half minute in length, the middle of the note being deafeningly shrill, loud and piercing to the ear, and its termination gradually lowered till the sound expires" (*1st Rept. Ins. New York*, p. 42).

They appear to be capable of producing different sounds, for Prof. Riley states that "when disturbed the noise they make mimics a nest of young snakes or young birds under similar circumstances — a sort of a scream. They can also produce a chirp somewhat like that of a cricket" (*1st Rep. Ins. Mo.*, p. 24).

Dr. Burnett has made the musical apparatus of the Cicada the subject of careful study. He found it to be wholly integumental in its nature, and not presenting any relation either by structure or analogy to the respiratory system. The drum is situated in each side between the thorax and abdomen, having its head, which is of the size of a marrow-fat pea, just under the point where the wings are attached to the body. It is a tense, dry, crisp membrane, crossed by cords or bars, produced by a thickening of the membrane, which meet on one side at the point of attachment of the muscles, which, by their contraction, keep it stretched. The sound is produced by a series of rapid undulations, running from the contracting muscles across the drum. The upper part of the abdomen serves as a sounding board, for with a portion removed, the sound is diminished in volume. A dry condition seems to be essential to the perfect action of the drum, as when it is moistened, or on wet days, the sound is very much diminished. The drumming is heard for four or five hours during the heat of the day, principally between the hours of twelve and two. In the female, there is no drum, nor any trace of the muscular apparatus belonging to it (*Proc. Bost. Soc. Nat. Hist.*, iv, 1851, p. 72).

Trees Attacked.

Dr. Harris mentions the oak as being their favorite tree in Massachusetts, and that they may be seen in oak forests in the middle of June in
such immense numbers that they bend and even break down the limbs
of the trees by their weight. Of fruit trees they prefer the apple.
Above fifteen hundred Cicadas have been found to have emerged from
the ground beneath a single apple-tree, leaving the surface of the ground
"as full of holes as a honey-comb." Other recorded food-plants are
locust, peach, pear, *Eupatorium*, hickory, chestnut, hazelnut, willow,
poplar, cottonwood, white cedar or arbor-vitæ (*Thuja occidentalis*), red
cedar (*Juniperus Virginiana*), and hemlock-spruce (*Abies Canadensis*)

The Cicada is not entirely dependent upon the roots of trees for its
food. Although confined to timber lands, and not occurring in treeless
regions or in those that have been under cultivation for seventeen years,
yet in an instance recorded in Michigan, where the timber had been
cleared for sixteen years and cultivated for most of that time, the Cicadas
came out of the ground the seventeenth year in the treeless fields quite
as numerously as among the timber. They appeared upon the same
day with the others, were as large, and to all appearance, equally well-
nourished, although for nearly their entire period, they must have sub-
sisted upon the roots of grasses and herbs.

Some of the facts connected with their appearance under strange conditions would indicate that their food-habits are yet far from being wholly
understood. In the instance recorded where they came up through the
ground floor of a cellar which had been dug to the depth of five feet,
it would be difficult to name the food — in kind and amount — that
could have supplied the requirements of the voracious larvæ when approaching their maturity.

Injuries to Vegetation.

The injuries to trees are the result of the countless punctures
made in the twigs for the deposit of the eggs. The twigs shrivel and
die soon after the hatching of the eggs, and eventually are broken off
by the winds, and fall to the ground, which is often covered by the dead
material. A writer mentions the appearance of the forests in Pennsylvania and Ohio for the distance of a hundred miles, just after the locusts had left them — looking as if they had been scorched by a fire
driving through them.

In the town of New Scotland, Albany county, N. Y., the Hudson
river valley brood was so abundant in the year 1826, that "they destroyed the fruitage of the orchards almost completely. Nearly all the

tender branches of the trees were so wounded in the deposit of the eggs that they broke from the main stems in the following year and fell to the ground, thus completely denuding the trees of their fruit-bearing branches " (Wm. G. Wayne, of Seneca Falls, N. Y., *in litt.*).

These injuries to fruit trees are often quite serious, especially when they are young. Almost every tree in young orchards has been known to have been killed by them. Peach and pear trees suffer severely from the punctures, and grapevines have also been badly injured.

The peach twigs sent from Canandaigua, in 1882, show unmistakably the severity of the attack. Pupa-cases were adhering to them, for the number emerging from the ground compelled their distribution not only over the trunk and principal branches but even their extreme tips before they could find the necessary space to attach themselves for their transformation. The fissures made in oviposition were very close — in some instances running into one another as a continuous slit for the extent of two and three inches. One piece of twig, but twelve inches in length, contained fifty-three of the fissures.

The punctures made by the insect with its beak, for food, also contributes to the death of the twigs. The female only is chargeable with this injury, as her longer life and the development of her eggs renders food essential to her. She has accordingly been provided with a complete digestive apparatus, while her consort, destined to live for but a few days, shows scarcely a trace of a digestive canal (Burnett, *Proc. Bost. Society Nat. Hist.*, iv, 1851, p. 71).

Natural Enemies and Checks.

When we recall the immense numbers in which these broods at times present themselves in their extension over hundreds of miles of territory, bending the branches beneath the weight of a half dozen or more individuals upon each leaf (Burnett, *loc. cit.*, p. 72), and endeavor to form some conception of the fearfully augmented successive broods that would result from an unchecked multiplication, we cannot but feel a profound gratitude for the beneficent operation of the conservative forces in nature, that effectually prevent such an undue increase. Of the five hundred eggs borne by each female, as a rule but two eggs only complete their cycle and develop into the perfect insect, to maintain the number of the preceding brood. Fortunate it is that the insects are exposed to so many casualties and to destruction from so many enemies. Dr. Harris remarks:

Their eggs are eaten by birds ; the young, when they first issue from the shell, are preyed upon by ants, which mount the trees to feed upon

23

them, or destroy them when they are about to enter the ground. Black-
birds eat them when turned up by the plow in fields, and hogs are
excessively fond of them, and, when suffered to go at large in the woods,
root them up and devour immense numbers just before the arrival of
the period of their final transformation, when they are lodged immediately under the surface of the ground. It is stated that many perish
in the egg state, by the rapid growth of the bark and wood, which closes
the perforations and buries the eggs before they have hatched,* and
many, without doubt, are killed by their perilous descent from the
trees.

Other insects also prey upon them. Dragon-flies (*Libellulidæ*) have
been seen to seize and devour them when they have just emerged from
their pupal state, and are still tender and helpless. The larvæ of some
species of fly, probably of the *Tachinidæ*, are known to feed upon them
internally.

It is a fortunate circumstance in the economy of this insect, that so
large a number of forest trees serve the purpose of oviposition equally
well with our fruit trees; were it otherwise, its attack would invariably
prove fatal to most of the young apple, pear and peach stock lying
within its range.

Preventives of Injuries.

No method is known by which these attacks may be prevented, and
very little can be done to mitigate them, when made by such innumerable hordes as are often witnessed. Something may be accomplished
by beating them or by picking them by hand from young trees in the
early morning and toward evening when they hang inactively upon the
twigs, and it would be practicable, by the aid of a strong wind blowing
in the right direction, to drive them with poles from the trees, compelling them to take wing and to seek the foliage of neighboring forest-
trees. Without the wind, they would probably, after a short erratic
flight, again return to the trees or transfer themselves to an adjoining
orchard.

Destroyed by a Fungus.

As the attack of the fungus, *Massospora cicadinæ*, referred to on
page 171, was observed by Prof. Peck to be of economic importance in
preventing the propagation of the species, his remarks thereupon are

---

* This sometimes occurs in thrifty young apple-trees, and in twigs of considerable size,
usually as large as a man's finger, where the injury caused by the deposit of the eggs
has not been sufficient to check, materially, the growth of the branch. In such cases the
twigs grow so rapidly, that in the course of the month which intervenes between the time
of the laying and the hatching of the eggs, the wound heals completely over, the tent or
splinters is nearly or quite overgrown, and the young insects never emerge from the eggs,
being inclosed in a living sepulcher (Le Baron, *Second Report on the Insects of Illinois*, 1872,
pp. 126-7).

herewith given, extracted from the 31st Report on the *N. Y. State Museum of Natural History*, 1879, pp. 19, 20:

It is a well-known fact that various insects are subject to the attack of parasitic fungi which prove fatal to them. The common house-fly is destroyed by one [*Entomophthora muscæ*], the silk-worm by another [*Botrytis Bassiana*], and the pupæ of different moths by others.

Another noticeable instance of this kind was observed the past season. It was found that the "seventeen-year locust," *Cicada septendecim*, which made its appearance in the Hudson river valley early in the summer, was affected by a fungus. The first specimen of this kind that I saw was taken in New Jersey, and sent to me by Rev. R. B. Post. Examination revealed the fact that the Cicadas or "seventeen-year locusts," in this vicinity, were also affected by it. The fungus develops itself in the abdomen of the insect, and consists almost wholly of a mass of pale yellowish or clay-colored spores which, to the naked eye, has the appearance of a lump of whitish clay. The insects attacked by it become sluggish and averse to flight, so that they can easily be taken by hand. After a time some of the posterior rings of the abdomen fall away, revealing the fungus within. Strange as it may seem, the insect may, and sometimes does, live for a time even in this condition. Though it is not killed at once, it is manifestly incapacitated for propagation, and, therefore, the fungus may be said to prevent, to some extent, the injury that would otherwise be done to the trees in the deposition of their eggs. For the same reason, the insects of the next generation must be less numerous than they would otherwise be so that the fungus may be regarded as a beneficial one. In Columbia county, the disease prevailed to a considerable extent. Along the line of the railroad between Catskill and Livingston stations many dead Cicadas were found, not a few of which were killed by the fungoid mass. As the insect makes its appearance only at intervals of seventeen years, and consequently will not be seen here again till 1894, it will scarcely be possible to make any further observations upon it for some time to come, yet it would be interesting to know how the fungus is propagated, or where its germs remain during the long interval between the appearance of the two generations. Do the fungus germs enter the ground in the body of the larvæ, and slowly develop with its growth, becoming mature when it is mature, or do they remain quiescent on or near the surface of the ground, waiting to enter the body of the pupa as it emerges seventeen years hence? Or, again, is it possible that the fungus is developed annually in some closely related species as the "harvest-fly," *Cicada canicularis*, and that it passes over from its usual habitat to the seventeen-year Cicada whenever it has the opportunity? These questions are merely suggestive; they cannot yet be answered.

Prof. Joseph Leidy had previously observed this fungus as early as in 1851 — see the *Proceedings of the Philadelphia Academy of Natural Sciences*, for 1851, v, p. 235 — but had not named it.

Chermes pinicorticis Fitch.

The Pine-bark Chermes.

(Ord. HEMIPTERA: Subord. HOMOPTERA : Fam. APHIDIDÆ.)

Coccus pinicorticis FITCH: in Trans. N. Y. St. Agricul. Soc. for 1854, xiv, 1855, pp 871-873; 1st Rept. Ins. N Y , 1856, pp. 167-169 (description of wing-less form).

Chermes pinifoliæ FITCH: in Trans. N. Y. St. Agricul. Soc. for 1857, xvii, 1858, p. 741; 4th Rept. Ins. N. Y., 1859, p. 55 (description of winged form).

Coccus pinicorticis. WALSH: in Pract. Entomol , i. 1866, p. 90 (the larvæ of *Aspidiotus pinifoliæ*).

Chermes pinicorticis. SHIMER: in Trans. Amer. Ent. Soc., ii, 1869, pp. 383-385 (winged form bred from the wingless).

Coccus pinicorticis. LINTNER. in Count. Gent. for Aug 21, 1873, xxxviii, p. 535.

Coccus pinicorticis. GLOVER. in Rept. Commis. Agricul. for 1876, p 44 (mention).

Chermes pinifoliæ. THOMAS. 8th Rept. Ins. Ill., 1879, p. 156 (from Fitch).

Chermes pinicorticis. OSBORN in Trans. Iowa St. Horticul. Soc. for 1878, xiii, p 400 (habits, etc., without name), id., for 1879, xiv, pp. 96-107 (diff. stages); Bull. Iowa Agricul Col., 1884, pp. 97-105 (general account).

Chermes pinifoliæ. PACKARD; Ins. Inj For.-Shade Trees, 1881, p. 118 (from Fitch).

The specific and common names of this insect indicate the tree upon which it occurs. Of the several varieties of pine, it is almost entirely confined to the Scotch pine (*Pinus sylvestris*) and the White pine (*Pinus strobus*). Its more usual habitat is upon the smooth bark around and below the axils where the limbs start from the trunk. Its appearance is that of little spots or patches of a white, flocculent, wool-like substance, adhering to the bark. This, however, is only a secretion, for the insect may be found beneath it when the covering matter is parted or brushed away.

The Secretion.

This white substance, which alone attracts the attention of the ordinary observer, as it completely hides the insect from view, is found, according to Dr. Shimer, most abundantly upon young white pines, near the ground in early spring, at about the time the frost is leaving the ground. Later it extends upward upon the trunk and into the axils of the branches. When examined with a magnifier, it has the appearance of " fine Saxony wool, the crinkled fibres drawing apart as do those of wool." Microscopic examination has shown it to be given out from some gland-like organs arranged in four rows upon the back and sides of the body of the insect.

Similar secretions are produced by many other of the *Aphididæ*. It occurs in the sub-family of *Aphidinæ*, in the beech aphis, *Phyllaphis fagi* (Linn.). In the *Pemphiginæ* we have many notable examples of it, as in the woolly aphis of the apple-tree, *Schizoneura lanigera* (Haus.); the woolly aphis of the elm, *S. ulmi* Thomas; the woolly aphis of the oak, *S. querci* Fitch; the alder blight, *S. tessellata* Fitch, etc. It also pertains to many of the species of *Pemphigus*, and indeed, according to Prof. Riley, is a feature common to the entire sub-family of *Pemphiginæ*.

From its occurrence upon so many species nourished by so greatly differing food-plants, it is not strange that in appearance it should be compared with such different substances as cotton, wool, silk, wax and meal. Some of the material examined microscopically by Buckton was found to consist of long flattened threads or fibres, with obscure, transverse and longitudinal striations, and when broken to show strong fractures as if they were brittle. That it was not of a waxy nature appeared from its insolubility in alcohol and a solution of potash — its scorching over a lamp without melting, and evolving a nitrogenous odor like that of burned feathers.

Of some of the species of *Pemphigus*, Dr. Thomas asserts that they do secrete a waxy substance that will melt under the heat of the hand.

For remarks upon the secretions of some other Hemipterous insects, pages 284–5 of my first report may be referred to.

The Insect and its Changes.

Within this secreted material, appearing as little downy balls fastened to the twigs and branches, upon its being pulled apart early in the spring the wingless mother of the pine-bark Chermes, may be found, together with a number of her small yellowish, moderately ovate eggs — perhaps from twenty-five to fifty.

Early in April, in favorable seasons, the eggs commence to hatch. The young larvæ emerge from the ball and travel actively over the bark for a short time. They are so small as to be almost invisible to the naked eye, of an oval outline, flattened, of a light brown color, with short legs, three-jointed antennæ, a sucking-tube extended beneath their body, and a short hair on each side of the abdominal segments. The two left-hand figures show the larva as seen from above and from beneath.

In the fore part of May, the larvæ are more abundant than at any other time. Their traveling soon ceases, and they attach themselves by their beaks to the tender bark of the young twigs. As their feeding proceeds, they increase rapidly in size, assuming a dark reddish-brown

color approaching black, while the secretion from their bodies commences, and soon hides them entirely from sight.

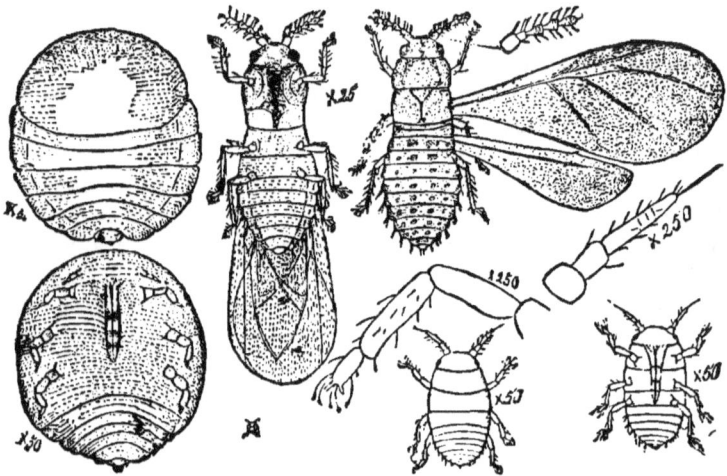

FIG. 48.— The Pine-leaf Chermes, CHERMES PINICORTICIS — the larva, the wingless female, and the winged form (male ?).

The time required to attain the mature form has not been recorded,
but the change probably occurs toward the last of May. The female destined to continue the species by laying eggs for another brood may be
recognized by being without wings, of an oval form which becomes
pyriform during egg-laying, with legs proportionally longer than in the
larval stage (see the enlarged view), and three-jointed antennæ. The lower
right-hand figures give upper and lower views of the female. From her
oviposition, a second brood results, which is followed by others, at unknown intervals until some time into the autumn.

In one particular, the first brood differs from the later ones, in that
it embraces a winged form. During the latter part of May, individuals
are to be found showing the pupal stage, by being provided with wing-
pads — otherwise they resemble the larvæ. The thorax is reddish, the
abdomen of a darker red, the wing-pads yellowish, and a little of the
woolly material adheres to portions of the body.

They continue in this stage but a few days, until about the first of
June, when their outer encasement parts asunder, and the winged form
issues, the appearance of which is shown in the figure. The antennæ
of these are five-jointed, the body is almost black, and the four wings

are white, transparent, and folded roof-like over the body.* The sex
of this form, or whether both sexes are embraced under it, seems not
yet to have been determined. They have not been observed in copulation or in the deposit of eggs. There is abundant room for careful
study in the life-history of this species, such as has been given to closely
allied forms, in the admirable paper of Messrs. Riley and Monell, giving biological notes on the Pemphiginæ.†

The above natural history of the species is mainly from observations
made upon it during the years 1878 and 1879, by Prof. Herbert Osborn,
of the Iowa State Agricultural College, to whom also we are indebted
for the figure used in illustration.

History.

An unfortunate mistake was made by Dr. Fitch in describing this
insect, in his First and Fourth Reports (loc. cit.), as two distinct species.
The single-jointed tarsus, the simple organization, and the waxy covering, led him to regard it as a *Coccus*, his first published observations
upon it not having been continued long enough to show him its subsequent development into the winged stage. When in after years
the winged individuals were met with, their Aphidian character was
at once recognized and they were correctly referred to the genus
Chermes, but without associating them with the wingless forms previously observed. And again in this stage, his usual careful habit of
observation is not apparent, for although the life-history of "the first
true Chermes that has been observed in this country," should have
entitled it to careful study at his hands, the following statement is quite
at variance with the later observations of others:

The females [winged] do not extrude their eggs, but clinging closely
to the leaf with their heads toward its base, they die, their distended
abdomens appearing like a little bag filled with eggs. The outer skin
of the abdomen soon perishes and disappears, leaving the mass of eggs
adhering to the side of the leaf, but completely covered over and protected by the closed wings of the dead fly.‡

* Prof. Osborn, who has apparently made a careful study of the venation of this insect,
states that "the posterior wings have a subcostal vein with *no branch veins*." A different
statement was made by Dr. Fitch in that the longitudinal rib-vein of the hind wings
"sends off, forward of its middle, a branch almost transversely inward, its tip curved
backward." Dr. Shimer corroborates Dr. Fitch, in stating from his examination of living
specimens that "the posterior wing has only a rib vein which forks opposite the quite conspicuous hook."

† *Bulletin of the U. S. Geological and Geographical Survey of the Territories*, v, 1879,
pp. 1–17.

‡ See above, from the observations of Prof. Osborn, the occurrence of *wingless* females
with their extruded eggs [and pyriform body during egg-laying], within little downy balls
fastened to the bark. Of *Chermes laricis*, of Europe, it is stated that as the apterous female
slowly deposits her eggs, she partly covers them with the down that she strips from her
body, and that they are piled around and upon her until she is nearly half buried in the
mass.

A strange blunder was made by Mr. Walsh in regarding the species
as identical with the pine-leaf scale insect (*Chionaspis pinifoliæ*).   He
wrote as follows to a correspondent : " The insects infesting the white
pine do not belong to the Aphis family (plant-lice), but to the Coccus
family (bark-lice).   The elongate white scale on the leaf was described
by Fitch as *Aspidiotus pinifoliæ ;* the downy patches on the bark as
Coccus pinicorticis.   But I believe that they are the same species, the
former containing the eggs, like the scale of the common bark-louse of
the apple-tree, and the latter being the young larvæ with downy matter
exuding from them."

In 1869, Dr. Shimer published the results of his study upon this insect,
called by him the " white-pine louse," made by him during the preceding year.   From specimens of *Coccus pinicorticis* Fitch, inclosed in a
feeding cage on June 3, the following day he obtained *Chermes pinifoliæ*
of Fitch (the winged form), thus clearly establishing their identity.

Four years later, forgetting the above article of Dr. Shimer, and following the lead of Dr. Fitch, I committed the error of referring to this
species as *Coccus pinicorticis,* in a notice of it contributed to the
Country Gentleman (see citation).

Other Species of Chermes.

The species of this genus are not numerous.   From this, in part, but
mainly for the reason, it may be supposed, that their oviporous reproduction limits their increase and consequent destructiveness much below
that of others of the family which multiply with such fearful rapidity
through the simple process of gemmation — they have failed to receive
the study bestowed upon their near relatives, the *Aphidinæ* and the
Pemphiginæ.

Two other pine-feeding species of *Chermes* occur in Europe, which
have not yet been detected in this country, viz., *Chermes pini* Koch,
and *Chermes corticalis* Kaltenbach.

Another species has been discovered by Dr. Fitch, in Washington
county, New York, extracting the juices from the leaves of the larch,
Larix Americana.   It was briefly characterized, mainly from wing
features, in the *Transactions of the N. Y. State Agricultural Society,*
for 1857, p. 753, and also appearing in the 4th *Annual Report on the
Insects of New York,* p. 66 (3d-5th Reports, 1859), as *Chermes larici-
foliæ.*

Dr. Packard in his *Guide to the Study of Insects,* 1869, p. 522-3, figures a pupal and a winged form resembling European species of *Adelges*
Vallot (*A. coccineus* of Ratzburg and *A. strobilobius* of Kaltenbach),
found by him in abundance, upon spruces in Maine, and producing

swellings at the ends of the twigs, resembling in size and form the cones of the same tree. Dr. Thomas, in his 8th *Report on the Insects of Illinois,* 1879, pp. 156–7 has given to this form the provisional name of *Chermes abieticolens.* The figure of Dr. Packard is introduced here in illustration of the genus.

In *Insects Injurious to Forest and Shade Trees,* 1881, p. 234, Dr. Packard, reproduces his remarks and figures from *Guide,* etc.,

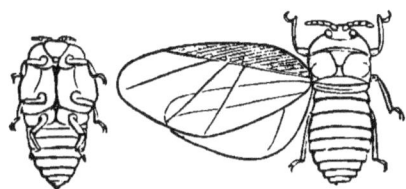

FIG. 49.— CHERMES ABIETICOLENS Thomas.

upon " *Adelges abieticolens* Thomas." He also records (p. 235) his observation of "*Adelges abietis* Linn." (*Chermes abietis*) of Europe in considerable numbers upon Norway spruces, in Salem, Mass.

A species quite injurious to the larch in Europe is *Chermes laricis* Hartig.

Another species depredating upon spruces in Europe has been described by Kaltenbach as *Chermes strobilobius.*

It is worthy of note that all of the above-named species—eight in number—are confined to coniferous trees. A single species only, so far as known to me, is found upon other vegetation, viz., the *Chermes atratus* of Buckton, occurring upon oak (*Quercus*).

Injuries of the Insect.

A tree badly infested with this insect becomes sickly and presents a slender, dwindled appearance; its leaves are short and stinted in their growth, and of a dull green color, and the annual growth of the tree is much curtailed (Fitch).

Its attack seems usually to be made upon trees that have been transplanted for the ornamentation of grounds, and is rarely observed upon those that are growing under their native condition.

According to Prof. Osborn, the Scotch pines in Ames, Iowa, and on the college grounds, were seriously infested for successive years.

At Fort Dodge, Iowa, the insect was reported as extremely abundant, and was observed literally piled up in the new growth.

Quite a severe attack was brought to my notice a few years since, through a communication from Tivoli, Dutchess county, N. Y., to this effect:

I inclose to your address a portion of a pine limb covered with a white substance, with the desire that you give me your opinion of what it is, and what I can do to remove it. The trees are large pines in the avenue, and are looking very badly — in fact, are slowly dying. It does

24

not affect all the trees, and those that are free from it are vigorous and healthy, which leads me to the opinion that this white substance is the cause.

The piece of limb sent was ten inches in length, nearly two inches in diameter, and was almost entirely covered with the insect and its flocculent mantle, some of the fibres of which were a quarter of an inch long. When received, August 7th, large numbers of the young, of almost an orange color and of an elongate form, were associated with the dark brown adults, showing that the attack of which complaint was made, was being actively and persistently continued.

Natural Enemies.

The multiplication of this insect is checked by quite a number of natural enemies that are known to prey upon it, and to be very active in its destruction.

Five species of lady-birds (*Coccinellidæ*) have been seen feeding upon it, viz., the Painted lady-bird, *Harmonia picta* Rand., Fig. 50; the Twice-stabbed lady-bird, *Chilocorus bivulnerus* Muls., Fig. 51; *Pentilia misella* Le Conte; *Scymnus terminatus* Say, and another *Scymnus* larva, differing from *S. terminatus*, but not reared to its perfect stage.

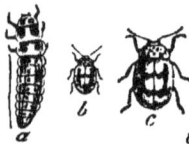

Fig. 50.—The Painted lady-bird, HARMONIA PICTA: *a*, the larva, enlarged; *b*, the beetle natural size; *c*, the same enlarged.

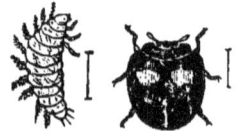

Fig. 51. —The Twice-stabbed lady-bird, CHILOCORUS BIVULNERUS: the larva and imago.

A larva of a lace-winged fly, *Chrysopa* species, that covered its back with the woolly material stripped from its victims, was observed by Dr. Shimer; also the larvæ of unknown species of Syrphus flies. These larvæ are known to render most excellent service in the destruction of plant-lice. Fig. 52 represents the larva of *Chrysopa*, and Fig. 53 that of a Syrphus fly in the act of devouring an *Aphis.*

Fig. 52.— Larva of a lace-winged fly, CHRYSOPA, sp.

Fig. 53.— Larva of a Syrphus fly.

The larva of a Heteropterous insect, resembling a small ant, was seen running actively about the infested pines, seizing and sucking the juices of the insects. It was determined by Mr. Uhler as *Camaronotus fraternus.**

Prof. Osborn discovered some round, black, hard and shiny mites within the downy balls of the *Chermes* which were believed to be feeding upon them. They evidently belonged to the genus *Oribates.*

---

* The resemblance of the European *Camaronotus cinnamopterus* Kirsch., to the large wood ant, *Formica rufa*, has been noted.

Remedies.

When the insect is found upon young trees the bark of which has not
become very rough, or when confined to the smoother bark of the
branches and twigs, most of them can be easily killed by going over the
bark and crushing them by the vigorous use of a stiff bristle brush.   In
connection with the brush a strip of cloth could be advantageously used
by applying it to the axils of the branches and drawing it repeatedly
backward and forward through them; or with a single turn around a
branch, moving it gradually outward upon the limb with the same saw-
ing motion.

Where larger trees and more extended surfaces are infested, a solution of whale-oil soap — a quarter of a pound to a gallon of water —
applied with a force pump, would destroy the insect.   The stronger application of kerosene emulsions required for adult scale-insects, would
not be needed for this Aphid, for its flocculent covering instead of serving as a protection to it only aids the efficacy of the applied insecticide
in absorbing and holding it until it may have its desired effect.

A method employed against the Larch Chermes of Europe, *Chermes
laricis* Hartig, which seems also to serve as a preventive of subsequent
attacks, is given by Miss Ormerod in her *Manual of Injurious Insects*
(p. 194-5):

To thirty-six gallons of water add half a pound of perchloride of
mercury; with this the infested trees are drenched in the early summer,
when the sap is flowing freely; a dry day is preferred for the operation, as it gives time for the solution to soak thoroughly into the bark.
This has been applied to ornamental trees and it is noted that trees
operated on in 1873, continued, at the time of writing (1880), free from
the "bug," and in thriving condition.   This application requires to be
in careful hands, being poisonous.   Woodpeckers that fed upon the
poisoned insects were destroyed by it.

Chimarocephala viridifasciata (DeGeer).

The Green-striped Locust.

(Ord. ORTHOPTERA: Fam. ACRIDIDÆ.)

Acridium viridifasciatum DE GEER: Mem. Hist. Ins., iii, 1773, p. 498, pl. 42, f.
    6; GOEZE: Gesch. Ins., iii, 1780, p. 325, pl. 42, f. 6.
Gryllus Virginianus FABR.: Syst. Ent., 1775, p. 291; Spec. Ins., i, 1781, p. 368;
    Ent. Syst., ii, 1793, p. 57.
Gryllus Locusta chrysomelas GMELIN: Linn., Syst. Nat. i, pt. iv, 1788, 2086.
Acridium Virginianum OLIVIER: in Encyc. Method., vi. 1791, p. 224.
Acridium marginatum OLIVIER: in Encyc. Method., vi., 1791, p. 229.
Acridium hemipterum PAL. DE BEAUVOIS: Ins. Afr.-Amer., 1805-21, p. 145, pl. 4,
    f. 3.

Œdipoda Virginiana BURMEISTER: Handb. d. Entomol., ii, 1838, p. 645.
Locusta (Tragocephala) viridi-fasciata HARRIS: Rept Ins. Mass., 1841, p. 147; Ins. N. Engl., 1852, pp. 158-9; Ins. Inj. Veg., 1862, p. 182, pl. 3, f. 2.
Locusta (Tragocephala) infuscata HARRIS: Rept. Ins. Mass., p. 147; Ins. N. Engl., 1852, p. 158; Ins. Inj. Veg., 1862, pp. 181-2.
Locusta (Tragocephala) radiata HARRIS: Rept. Ins. Mass., 1841, p. 148; Ins. N. Engl., 1852, p. 159; Ins. Inj. Veg., 1862, pp. 183-4.
Gomphocerus viridi-fasciata UHLER: in Harr. Ins. Inj. Veg 1862, p. 181; ib. as *Gomphocerus infuscatus* and *G. radiatus.*
Tragocephala viridi-fasciata. SCUDDER: in Bost. Journ. Nat. Hist., vii, 1862, p. 461; Cat. Orthop. N A.—Smithson. Miss. Coll, No. 189, 1868, p. 82; Entomolog. Notes, iv, 1875, pp. 80-82.
Tragocephala viridifasciata. GLOVER: Illus. N. A. Ent.—Orthop., 1872, pl. 5, f. 9.
Tragocephala viridifasciata. RILEY: 8th Rept. Ins. Mo., 1876, p. 149, f. 46.
Chimarocephala viridifasciata SCUDDER: in U. S. Geograph. Surv. West of 100th Mer., 1876, p. 508.
Tragocephala viridifasciata. THOMAS: 9th Rept. Ins. Ill., 1880, pp. 85, 93, 105-6, figs. 13, 17: Synop. Acrid. N. A. (in Hayden's Rept. U. S. Geolog. Surv. Terr., v, 1873), 1873, pp. 103-4, pl. f. 3; id., p. 102, as *T. unifasciata* Harr.
Tragocephala viridifasciata. LINTN.: in Count. Gent., Mch. 9, 1882, xlvii, p. 189 (winter appearance).
Chamarocephala viridifasciata. BRUNER: in 3d Rept. U. S. Ent. Commis., 1883, p. 56 (in list of N. American Acrididæ).

Midwinter Appearance of the Insect.

In several portions of the State of New York — at its eastern and western extremities, and at intermediate localities, much interest was excited, and in some cases considerable alarm, by the sudden appearance in the pastures and meadows at about the middle of February, 1882, of large numbers of this insect, hopping about in the warm sunshine, almost as actively as in the month of June.

My attention was called to them by a note received from Mr. G. M. Gillette, of Bergen, Genesee county, in which he wrote:

I to-day mail you what I should term grasshoppers if it were only later in the season. They are as plentiful as in the month of July. On the 14th of this month, they could be seen by the million sunning themselves on the south side of stone walls in this vicinity.

Examination showed them to be the young of the Green-striped locust, the *Locusta (Tragocephala) viridifasciata* of the Harris Reports.

They measured from four-tenths to six-tenths of an inch in length — the smaller being in the larval stage, without indication of the wing-pads, and the larger in the pupal stage,

FIG 51.—The Green-striped Locust, CHIMAROCEPHALA VIRIDI-FASCIATA: *a*, larva; *b*, adult.

with wing-pads extending half the length of the abdomen. In Fig. 54, the larva is shown at *a*, and the full-grown and winged insect at *b*.

Upon addressing Mr. Gillette for further information concerning them, and proposing some observations desirable to be made, the following reply was received under date of February 28th.

Their Appearance in Genesee County.

The young grasshoppers were first seen as early as the 10th of February. No winged ones were observed, and 1 think it is safe to say that there were none. When first noticed, they were upon an old stone-wall, apparently sunning themselves.

On February 14th, having occasion to cross an old unused meadow, and also a ten-acre lot of fall-sown wheat, I found them in great numbers. They were very active, and it was no little effort to catch and cage those sent to you. There was no snow upon the fields at the time, but along the wall where there were patches of snow, they were as lively as upon the grass, and seemed rather to enjoy a good "hop, skip and jump" upon the smooth surface. There have been no birds in this vicinity [reply to question], consequently I do not think that they have been disturbed. They may have been attacked by snow-birds, although quite scarce, but not to my knowledge.

Later, upon the 1st of April, Mr. Gillette saw "a large flock of black-birds [probably the red-winged black-bird, *Agelæus Phœniceus*] feeding sumptuously upon the poor persecuted grasshoppers. The birds came in immense flocks." The grasshoppers had not up to that date increased perceptibly in number or in size. The nights had been cold, "but the 'hopper was early up waiting for the sun." Of a half-dozen specimens inclosed in a box February 15, without food, three were alive two weeks later. Upon the 22d of April, Mr. G. wrote: "I have yet one grasshopper alive of the original six. He has to-day completed his ninth week without a morsel of food, and although he seems to be failing, I hope to keep him another week and give him a good record." — "The sole survivor made his escape after a confinement of ten weeks and three days."

The grasshoppers continued to be seen at Bergen until April 9th after having been observed for nearly a month.

Their Appearance in Westchester County.

The Albany *Evening Journal*, of February 21st, contained the following notice:

The farmers in the vicinity of Mount Kisco, Westchester county, N. Y., are greatly troubled in mind concerning the appearance of a black grasshopper which was noticed two or three days ago in the fields. They were first discovered by George Sypher on the farm of Isaac Thorne. The young man was walking through the field, and he noticed

•

the insects skylarking on the beautiful snow.   Having never seen such
a sight in midwinter, Sypher was at first inclined to take to his heels
and run for his life, fearing that something had got wrong in his head.
But finding that he had full control of his senses, he began an investigation of the strange phenomenon, and when he had satisfied himself
that there was no fiction about the grasshoppers, he caught half a dozen
of them and introduced them to Mr. Thorne.   The gentleman was also
surprised, and immediately started for the field to carry on a more
thorough investigation.   There he found scores of the insects, skipping
around in the sunshine, and after securing a bottle full, went among his
neighbors exhibiting them.   They were all greatly astonished at the
hoppers' appearance, declaring that such an invasion was never heard of
before in the winter in that region.

Since their appearance on Mr. Thorne's place, they have been reported in a number of localities in that vicinity.   The farmers express
alarm lest a new pest should make its appearance next summer, of
which these hoppers are the advanced guard.   The insects are jet
black [?] and somewhat smaller than a cricket.   Nothing like them was
ever seen before at any time.

No Cause for the Alarm.

Other notices of the general character of the above appearing in several of the newspapers of the State, and my attention being called to
them, — in a communication to the Albany *Evening Journal* of February 25th, I gave the assurance, that although these grasshoppers had appeared " by millions," yet the farmers need entertain no apprehension
that their occurrence at this time indicated extensive depredations from
them the ensuing season.   It was not unusual for specimens of the species in moderate numbers to emerge from their winter retreats in stone-
walls and elsewhere during a continuance of warm weather in winter.
They had never been known as a particularly injurious species, as the
red-legged grasshopper, *Caloptenus femur-rubrum*, at times, proves to
be.   Their appearance in immense numbers at this time should be regarded as a favorable omen, as promising diminished numbers during
the summer months, for very many of those which had been prematurely aroused from their winter sleep would fall victims to the cold
which sharply followed their awakening.

The Attendant Meteorological Conditions.

As will be more fully stated hereafter, it is the habit of this species
to pass the winter in an inactive state, in the form of larvæ and pupæ,
in sheltered places in old stone-walls, beneath roots of grasses, under
piles of rubbish, etc.   Under the ordinary temperature of the winter
months, their quiet hibernation would not be disturbed, nor their winter's sleep broken until the warm days of early May aroused them from

their torpor, and summoned them to activity and to the food awaiting
them in the already green fields.   But at the time above stated for the
premature awakening and coming forth of *C. viridifasciata*, we find
from the records that unusually warm weather was prevailing.   From an
average temperature for the several preceding weeks of 27° Fahr., it
suddenly changed to a mean temperature (of one week) of 40°, reaching at the highest 56°.   The records of the U. S. A. Signal Service, at
Albany, show the following temperatures for the months of January and
February, of 1882:

January   1–7, mean temp., 23°.   Highest temp., 41°.
  "   8–14,   "   37°.   "   47°.
  "   15–21,   "   30°.   "   48°.
  "   22–28,   "   21°.   "   40°.
  "   29–Feb. 4, "   25°.   "   45°.
Feb.   5–11,   "   33°.   "   47°.
  "   12–18,   "   40°.   "   56 .
  "   19–25,   "   29°.   "   40°.
  "   26–Mch. 4, "   41°.   "   56 .

The month of February is recorded as "much the warmest February
since the establishment of the Signal Station " at Albany, in 1873, the
mean temperature of the 16th having been 50°.   The December preceding had also been the warmest during the same period, having had
a mean temperature of 39 , as against an average of 27° for the seven
previous years.

Description of the Insect.

Dr. Harris gives the following description of the species: Green;
thorax keeled above; wing-covers with a broad green stripe on the outer
margin extending from the base beyond the middle and including two
small dusky spots on the edge, the remainder dusky but semi-transpar-
ent at the end; wings transparent, very pale greenish yellow next to the
body, with a large dusky cloud near the middle of the hind margin,
and a black line on the front margin; antennæ, fore and middle legs
reddish; hind thighs green, with two black spots in the furrow beneath;
hind shanks blue-gray, with a broad whitish ring below the knees, and
the spines whitish, tipped with black.   Length about one inch; expanse
from more than one inch and three-fourths to nearly two inches.

Dimorphic Forms.

The species, according to Mr. Scudder, is quite variable, but presents
an interesting case of dimorphism, in that it appears under two distinct

forms. For the one, the form above described, he has retained the name under which it had been described by Fabricius, viz., *Virginiana* (see the Synonymy given), and for the other, *infuscata* — the name under which it had been described by Dr. Harris as a separate species. The two differ in the presence or the absence of bright green colors, which in the former replace the gray of the latter, in the whole of the head, the front part of the thorax (pronotum), the hind thighs, and on the greater portion of the costo-basal half of the fore-wings, and in spots beyond the middle of their front border. These differences appear to be mainly sexual, for in about one hundred and fifty examples examined by Mr. Scudder, 84 per cent of the males were *infuscata*, and 77 per cent of the females *Virginiana*.

Synonymy.

It will be seen from the table of synonymy given, that the species has been peculiarly unfortunate in the number of names that it has received both generic and specific — Dr. Harris having made three species of it.

The generic name under which it is herein included, was proposed for it by Mr. Scudder in consideration of *Tragocephala* being preoccupied in the Coleoptera.

Habits and Natural History.

As soon as the weather becomes sufficiently warm in the spring, these young grasshoppers come from their retreats and commence to feed. Many of them (the more advanced) have attained their maturity and acquired their wings by the middle of May. A few days after maturity the female deposits her eggs. By means of the horny appendages with which the tip of her abdomen is provided, she drills a hole in the ground the length of her abdomen, which she proceeds to fill with from twenty to thirty eggs arranged symmetrically in four rows, and forming, when the operation is completed, a pod-like mass, firmly cemented together by a mucous matter secreted with the deposit of each separate egg. Fig. 55 shows the oviposition of the Rocky Mountain locust, and as

FIG. 55. — *a, a, a,* Female in different positions ovipositing; *b,* egg-pod extracted from the ground, with the end broken open; *c,* some separate eggs; *d, e,* a section showing an egg-pod placed and another being placed; *f,* where a pod has been covered up.

the methods are very similar in the *Acrididæ* it may illustrate that of this species. Two or three weeks later, another deposit of eggs is made in the same manner, and a third, and perhaps a fourth, at intervals thereafter.* The eggs hatch in about three weeks, and produce the forms mentioned by Dr. Harris as occurring " in various states of maturity, in pastures and mowing lands, from the first of June to the middle of August," in the New England States.

After leaving the egg, the insect undergoes at intervals two moltings without showing the wing-pads that are later developed. These are known as the three larval stages. Another molt brings it to its first pupal stage, in which the small wing-pads are in position, turned upward over the back. The fourth molt develops the fifth stage, or the true pupa in which the wing-pads attain their greatest development and the thorax has assumed nearly the form of that of the mature insect. With the fifth molting the mature, winged insect appears. Fig. 56 represents the larvæ at different stages and the true pupa of the Rocky mountain locust.

FIG. 56.—CALOPTENUS SPRETUS: a, a, newly-hatched larva; b, full grown larva; c, the pupa.

The molting operation. — The moltings of insects are so wonderful, and so little understood, and those of the grasshoppers are so very rarely seen — not once where a thousand are observed of the butterflies and moths — that I borrow the excellent illustrations and explanations of Prof. Riley of the final molting of the Rocky mountain locust, from the 1st volume of the *Report of the U. S. Entomological Commission.* Those who have not access to this volume, or who, among so much other highly interesting material, may have overlooked this portion, will be grateful for this reproduction:

When about to acquire wings, the pupa crawls up some post, weed, grass-stalk or other object, and clutches such object securely with its hind feet, which are drawn up under the body. In doing so, the favorite position is with the head downward, though this is by no means essential. Remaining motionless in this position for several hours, with antennæ drawn down over the face, and the whole aspect betokening helplessness, the thorax, especially between the wing-pads, is noticed to swell. Presently the skin along this swollen portion splits right along the middle of the head and thorax, starting by a transverse curved suture between the eyes and ending at the base of the abdomen.

Let us now imagine that we are watching one from the moment of this splitting, and when it presents the appearance of *a* in Fig. 57. As soon

---

* *Caloptenus femur-rubrum* has been observed by Prof. Riley to deposit four "egg-pods," averaging one hundred and four eggs each, within a period of sixty days, giving an egg-laying period for the species of about two months.

as the skin is split, the soft and white fore body swells and gradually extrudes more and more by a series of muscular contortions; the new head

Fig. 57. — The molting operation of a " grasshopper " (the Rocky mountain locust); *a*, pupa with skin just split on the back; *b*, the imago extruding; *c*, the imago nearly out; *d*, the imago with wings expanded; *e*, the imago with all parts perfect.

slowly emerges from the old skin, which, with its empty eyes, is worked back beneath, and the new feelers and the legs are being drawn from their casings and the future wings from their sheaths. At the end of six or seven minutes, our locust — no longer pupa and not yet imago — looks as at *b*, the four front pupa legs being generally detached and the insect hanging by the hooks of the hind feet, which were anchored while yet it had that command over them which it has now lost. The receding skin is transparent and loosened, especially from the extremities. In six or seven minutes more of arduous labor, of swelling and contracting — with an occasional brief respite, the antennæ and the four front legs are freed, and the fulled and crimped wings extricated. The soft front legs rapidly stiffen, and, holding to its support as well as may be with these, the nascent locust employs whatever muscular force it is capable of to draw out the end of the abdomen and its long legs, as at *c*. This in a few more minutes it finally does, and with gait as unsteady as that of a new-dropped colt, it turns round and clambers up the side of the shrunken, cast-off skin, and there rests while the wings expand and every part of the body hardens and gains strength — the crooked limbs straightening and the wings unfolding and expanding like the petals of some pale flower. The front wings are at first rolled longitudinally to a point, and as they expand and unroll the hind wings, which are gathered and tucked along the veins, at first curl over them. In ten or fifteen minutes from the time of extrication these wings are fully expanded and hang down like dampened rags (*d*). From this point on, the broad hind wings begin to fold up like fans between the narrower front ones, and in another ten minutes they have assumed the normal attitude of rest. Meanwhile the pale colors which always belong to the insect while molting have been gradually giving way to the natural tints, and at this stage our new-fledged locust presents an aspect fresh and bright (*e*). If now we examine the cast-off skin, we shall find every part entire with the exception of the rupture which originally took place on the back; and it would puzzle one who had not witnessed the operation to divine how the now stiff hind shanks of the ma-

ture insect had been extricated from the bent skeleton left behind.
They were in fact drawn over the bent knee-joint, so that during the
process they were doubled throughout their entire length. They were
as supple at the time as an oil-soaked string, and for some time after
extrication they show the effects of this severe bending by their curved
appearance.

The molting, from the bursting of the pupa-skin to the full adjustment of the wings and the straightening of the legs of the perfect insect,
occupies less than three-quarters of an hour, and sometimes but half an
hour. It takes place more frequently during the warmer part of the
morning, and within an hour after the wings are once in position, the
parts have become sufficiently dry and stiffened to enable the insect to
move about with ease; and in another hour, with appetite sharpened by
long fast, it joins its voracious comrades and tries its new jaws.

Nearly all of the first brood of the hibernating individuals have
passed away by the first of August. Eggs had been deposited at various times by those from about the first of June into July. The first
larvæ of the second brood may be seen abroad as early as the latter
part of June, and continue to appear throughout the following month.
The earliest of these maturing and becoming winged about the 10th of
August, would toward the latter part of the month deposit eggs for the
following brood. Oviposition continues through September, and in
September also, young larvæ are again seen. These, from the lower
temperature of the advanced season, mature but slowly, and are destined
to pass the winter in their immature forms, some of them having pro-
gressed to their first pupal stage. They have been observed in Iowa, in
very large numbers, abroad in October and November. Mr. A. H.
Gleason, of Little Sioux, Iowa (N. Lat. 41° 50′, and Isotherm of 48°) has
written as follows of them:

They lay their eggs in August and September, and these hatch (at
least some of them) the same fall. I saw them last October and November, little fellows in spots of from one square yard to a twenty acre
piece covering the ground as thick as ever I have seen the Western
plague [*Caloptenus spretus*]. I have found them in the winter, under
the leaves and dry straw and husks that have drifted up under the
fences and behind logs in the woods, in a dormant state, and upon warm-
ing them they would become brisk as ever (*First Report U. S. Entomolog. Commis.*, 1878, p. 459).

A Double-brooded Species.

It will be observed that this species is an exception to most of the
Acrididæ in its having two broods a year. That it is double-brooded
has not been previously published, but it follows from the observations
above recorded, and it is corroborated by the following notes made by
Prof. Riley upon the species in Missouri, which he has kindly sent me,
for present use:

"Larvæ in different stages, pupæ and winged insects were found May 2d. The winged specimens had not hibernated but matured during April. The winged insects were found till about the 15th of June.

"The first larvæ were noticed June 26th, although there may have been some earlier. The eggs hatch in about fifteen days.

"Winged insects are again found in September, and larvæ in September and October. The latest deposited eggs winter over and hatch during March and April.

"This shows the species to be two-brooded. The first brood, from eggs deposited in the fall, some of which hatch, the larvæ wintering over, while others hatch in spring. Specimens acquire wings from the latter part of April until the middle of June.

"Eggs from the spring brood are deposited the latter part of May and early part of June, and the winged insects from these eggs are noticed in August and September."

Other Double-brooded Species.

It has been claimed by some writers that the Rocky mountain locust, *Caloptenus spretus*, produces two generations annually, but a careful examination of the evidence educed in favor of the claim shows it to be, according to Prof. Riley, "overwhelmingly in favor of normal single broodedness. * * * * * * * * While we admit the possibility of a second generation, we believe that it is exceptional, and that the insects composing such second generation seldom, if ever attain maturity or perpetuate their kind" (*First Rept. U. S. Entomolog Commis.*, 1878, pp. 242, 243).

Of *Caloptenus atlanis*, very nearly allied to *C. spretus*, Prof. Riley states (*loc. cit. sup.*), "we have proved it to be double-brooded." This statement should be qualified, by referring it to St. Louis and its more southern localities, for in a communication received from Prof. Riley, he expresses his doubt if, at St. Louis, the second brood is always successful in perpetuating itself, and adds that "later observations in New England convince me that in northern localities it is invariably single-brooded."

In addition to *C. viridifasciata*, we may expect that there will be found to be a spring and an autumn brood in at least several of the species of *Stenobothrus* and *Tettix*, some of which are known to hibernate in the half-grown condition. Of the former, *S. maculipennis* Scudder, shown in Fig. 58, occurs in the State of New York, together with *S.*

Fig. 58.— STENOBOTHRUS MACULIPENNIS —a, mature insect; b, pupa; c, larva.

FIG. 59. — TETTIX GRANULATA — the granulated Grouse locust

æqualis Scudd., *S. bilineatus* Scudd., *S. propinquus* Scudd., and *S. cur-*

tipennis (Harris). Of the latter, *Tettix cucullata* (Burm.), *T. triangularis* (Scudd.), *T. granulata* (Kirby), shown in Fig. 59, and *T. ornata* (Say) are also New York species. Dr. Harris seems to record two broods of the last-named species in his statement, that of his *Tetrix sordida* — a synonym of *T. ornata*, he had taken individuals both in May and September.

In all cases in which large numbers of a species of the *Acrididæ* are observed to hibernate in the larval stage, such hibernation may be regarded as presumptive evidence of two generations annually — of a spring and autumn brood.

Remedies.

For the remedial measures to be employed against this species when it occurs in unusual and destructive numbers, the agriculturist is referred to the *First Annual Report of the United States Entomological Commission*, where the various remedies and devices available against the Rocky mountain locust, are discussed at length, in chapter xiii of 70 pages.

[ADDITIONAL NOTE. — A remedy that gives promise of being a most excellent one against the grasshopper visitations that occur in the Middle and Eastern United States, has been published in the *Pacific Rural Press* of July 4, 1885. The results attending its use in the protection of vineyards and orchards in California have been of a very satisfactory character. If future experiments substantiate the claim made for its efficient operation, it will prove to be one of the most valuable insecticides thus far given to our fruit-growers and gardeners. It is as follows, as communicated by Mr. D. W. Coquillet, to whom appears to belong the credit of its discovery:

It consists of a mash composed of bran, arsenic, sugar and water, the proportions being one part of sugar, one and one-half parts of arsenic and four parts of bran, to which is added a sufficient quantity of water to make a wet mash. A common washtubful of this mash is sufficient for about five acres of grapevines. Fill the washtub about three-fourths full of bran, add six pounds of arsenic, and mix it thoroughly with the bran; put about four pounds of coarse brown sugar in a pail, fill the pail with water and stir until the greater part of the sugar is dissolved. Then pour this sugar-water into the bran and arsenic, and again fill the pail with water and proceed as before until all of the sugar in the pail has been dissolved and added to the bran. Now stir the latter thoroughly and add as much water as is necessary to thoroughly saturate the mixture, and it is ready for use.

Throw about a tablespoonful of this mixture upon the ground beneath each vine infested with grasshoppers; and in a short time the latter will leave the vine and collect upon the bran and soon commence feeding upon it. Those which are upon the ground six or eight feet from the bran will soon find their way to it, apparently guided by the sense of smell, as those to the leeward of the bran have been observed to come to it from a greater distance than those which were upon the side of the bran from which the wind was blowing. After eating as much of the bran as they desire, the grasshoppers usually crawl off, and many hide themselves beneath weeds, clods of earth, etc., and in a few hours will be found to be dead.

The mixture costs from thirty-five to forty cents per acre of vineyard, including labor of mixing and applying it. In orchards the cost will be considerably less than this. One man can apply it to eight or ten acres of vineyard in a day.

I have seen this remedy tried on an extensive scale at the vineyard and orchard of Messrs. Kohler, West and Minturn, at Minturn station, Fresno county. In that part of the vineyard which was the most thickly infested by grasshoppers from thirty to fifty dead grasshoppers were found beneath almost every vine, while beneath the adjacent weeds were hundreds of others, the greater part dead. It was also very effectual when placed beneath small fruit trees, the grasshoppers leaving the trees to feed upon this mixture.

The addition of sugar to this mixture is merely to cause the arsenic to adhere to the particles of bran, and not for the purpose of increasing its attractiveness, since it was found that the grasshoppers were not attracted to pure sugar. Middlings or shorts have been used in the place of bran, but are not so desirable, since in drying they assume a solid mass which the grasshoppers cannot eat, whereas bran in drying never assumes a solid form.— D. W. Coquillet, *Atwater*, *MercedCo.*, June 27.]

Atropos divinatoria (O. Fabr.).

(Ord. Neuroptera: Fam. Psocidæ.)

Termes divinatorium Müller: Zoöl. Dan. Prodr., 1776, p. 184, No. 2179.—O. Fabr: Fn. Grönl., p. 214, No. 181.—Linn.: Syst. Nat., Ed. Gmelin, p. 2914, No. 8.

Atropos divinatoria Hagen: Neurop. N. Amer., 1861, p. 8, No. 1; Ent. Month. Mag., ii, 1865, p. 121, No. 1; Stett. Ent. Zeit., xliv, 1883, pp. 289-293.—McLachl.: Month. Mag., iii, 1866, p. 180, f. 1.

Troctes divinatorius Kolbe: Psoc., p. 133, No. 1.—Provancher: Faune du Canada, ii, p. 66.

Troctes fatidicus Burm.: Man. Ent., 774, No. 2.

Liposcelis museorum Motsch.: Etud., i, p. 20.

The above synonymy of the species is from Dr. H. A. Hagen's *Monographic der Psociden* published in the *Stettiner Entomologische Zeitung* for July-September, 1883.

Remarkable Occurrence of the Insect in Beds.

Numerous examples of the insect were received from Otsego county, N. Y., with a request to return answer what they were and how to remove them from a room and bed where they abounded. It was thought that they had been left in the room by some traveling agents who had occupied it a year previous. Effort was made to exterminate the insect by hot water and corrosive sublimate. The means used seemed for a while to be successful, but the following year (in April) they again appeared in large numbers.

Answer was returned that they were a low form of the order of Neuroptera, known as *Psocidæ*, and bearing the scientific name above given.

It was further stated that the insect could not have been introduced in the manner supposed, viz., from the bodies or the clothing of the travelers, for the reason that, although in general appearance it is quite louse-like, yet it does not belong to the *Pediculidæ*, and is not parasitic upon the persons of men or animals. It frequently occurs in books and among old papers, and is known also to feed upon the remains of minute insects, and to attack the insects in entomological collections. Its mouth parts are not fitted for sucking blood but consist of strong cutting jaws.

Recommendations for its Destruction.

To remove the insects from the bed, it was recommended to wash the entire bedstead with hot water and soap-suds, or to brush over the entire surface with benzine, if it could be done without injury to it. Such portions of the bedding as could not be washed, as the pillows and mattress, should be exposed to the hot sunlight for two or three days, turning them from side to side occasionally. A feather-bed might require some benzine to be poured into it, and the contents of a straw tick should be burned.

If the room was infested to any considerable extent, if papered, the old paper should be removed, and the walls thoroughly washed before re-papering. Carpeting should be treated as recommended for the bedding, and the floors and the entire wood-work washed with strong soap-suds.

Particulars of the Invasion.

The absolute certainty expressed by the lady from whom the specimens were received, that they had been introduced into her house by the travelers above stated, and the certainty, from an entomological standpoint, that they could not have been thus introduced, led to considerable correspondence upon the subject, which elicited minute and extended particulars respecting the suddenness of the invasion, its abundance, its persistence, and fruitless efforts to conquer it. These proved to be of considerable interest in their economic aspect, as will be seen from the extracts upon these several points, that we make from the letters sent to me.

"The room had been newly papered two years ago, and an almost new rag carpet was on the floor. In two recesses were cord bedsteads which had never been used except as 'spare beds.' Last spring the room was cleaned, but the carpet was not taken up. The strawbeds were newly filled, the straw of which was all right [in answer to the suggestion that the straw might have been infested], for with three other beds filled with the same there has been no trouble. The floor underneath the bed is not carpeted. The foot-curtains had been ironed and put up the afternoon before the bed was occupied by the men. I know the beds and room were clean as could be, with no vermin present. I never keep any thing but a light spread on my spare beds. I made the bed with clean sheets and a comfortable and bed-quilt from the closet.

"The bed was occupied for the night by two young men who were taking orders for enlarging photographs and finishing in India ink. They had no luggage with them or any packages except the cases containing pictures.

"In the morning, about an hour after their departure, the bed was opened, and in the sheets were large vermin (lice) [these were not sent]. The sheets were carefully gathered up and brought down stairs, and out of doors. The bed-quilt, comfortable and bed-curtains were literally covered with the creatures such as I sent you.   *   *   *   The bed not slept in was worse, if possible, with these little specimens than the one occupied. We think the men must have distributed their garments around the room. That afternoon I washed the sheets, pillow cases, straw-ticks and curtains of both beds. We threw the bedding out of doors, and killed all that we could find, and then ironed them on both sides. The bedsteads we washed in hot water and with hot brine where they would not be injured by it, and the floor the same. While they were yet damp, we took Persian insect powder and puffed it into every crevice in the floor and bedstead.

"No [in reply to question], there are none elsewhere in the house, and never *there*, until these lodgers left them. We did not settle the room again for weeks, but kept up a diligent search, and every few days would find two or three. The last warm weather in the autumn, I found quite a number on the fine pillow cases next the ticks. I had the straw tick emptied and washed again, and ironed the bedding. I kept the window raised during most of the winter, hoping that the extreme cold would put an end to them. But to my surprise the first warm day this spring, I found the white spread on the bed where they slept almost covered again, from which I gathered those last sent to you. I never have found them larger [reply to question]. They do not soil a white spread when I kill them on it. They will be in clusters, frequently, of twenty or more. They appear to like white goods, cotton or linen. I have continued my warfare upon them faithfully while waiting to hear from you. I took up the carpet, but there was nothing in it or under it, not even dust, it was so clean. We uncorded the bedsteads and examined them, but there was nothing in them. I do not think that they incline to wood or wall.

"Gasoline is the most effectual of any thing that I have tried for killing

them.   One hundred dollars would be no temptation to me to pass
through such a trial again, even if I have succeeded now in subduing
them.   I had cared for the room twenty-three years, and never saw any
thing of the kind before.   It was not *so* when the men retired, and was
literally alive an hour after their departure.   I know that it seems incredible, but my testimony can be fully substantiated.   There are some
that *dart* so that I cannot trap them.   It seems to me that they are in
different stages of development, or perhaps different varieties."

The Possible Source of the Insects.

I know of no other instance in which *A. divinatoria* has occurred in
this country, in an abundance equal to the above, nor am I able to offer
any explanation for its sudden appear-
ance.   The only suggestion as to its
source that presents itself is this; its
possible introduction with the straw
used in filling the straw-ticks, notwithstanding the statement that other
beds filled with straw from the same
source were not infested.   Species of
Psocidæ are known to abound in
barns.   Dr. Hagen informs me that he
had, on one occasion, found more
than half of the refuse material left
in a barn after threshing the grain to
consist of a small species of Psocus.
Mr. McLachlan, of London, England,
has found " myriads " of this same species, *Atropos divinatoria*, in the straw
bottle envelopes in the wine cellar of

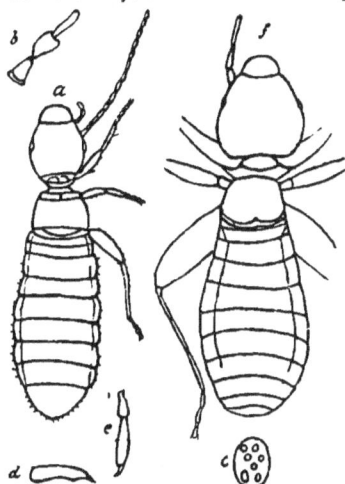

FIG. 60.—ATROPOS DIVINATORIA: *a*, var. *cucurbitæ*; *b*, side-view of basis of antenna; *c*, eyes;
d, claw; *e*, tarsus of nympha; *f*, *A. divinatoria*,
drawn from life.

his house, associated with *Clothilla picea*.   Examples of *Clothilla pulsatoria* (to be noticed hereafter) have been sent to me for name, taken
from cattle-stalls, in Warren, Ohio, where they were abounding, and
from the locality that they occupied, were supposed to be cattle-lice.

Description of the Insect.

The American forms of this species have been found by Dr. Hagen
to differ in several particulars from the European ones, but as these
differences are mainly colorational, they are regarded as identical.   They
have been minutely described by Dr. Hagen in "Beiträge zur Monographie der Psociden — Familie Atropina," in *Entomologische Zeitung,
entomologischen Vereine zu Stettin*, 44, 1883, pp. 285–332, and figured in
the same publication (Plate ii, Fig. 4) of the preceding year.   To these,

26

the student who may desire knowledge of the species, is referred. In lieu of description, the accompanying figures (Fig. 60) are presented, which will serve for the identification of the species. They are copied from the plate above referred to.

Habits of the Psocidæ.

Of the general habits of the *Psocidæ* Westwood remarks that they frequent the trunks of trees, palings, old walls, stones covered with lichens, old books, &c., for the purpose of feeding either upon the still more minute animalculæ which inhabit those situations, or, more probably, upon the decaying vegetable matter to be there met with. They are extremely active, and when approached, they endeavor to hide themselves by running to the opposite side of the tree, or other object on which they are stationed. The perfect insects are produced toward the end of summer, when they sometimes appear in great numbers. The larvæ and pupæ are equally active with the imago, from which the former differ in being apterous, while the pupæ have rudimental wings (Westw. *Classification of Insects*, ii, 1840, p. 18).

The Death-Watch.

An interesting member of the family is the *Clothilla pulsatoria*, of which a figure is presented from Packard's Guide, which may not, how-

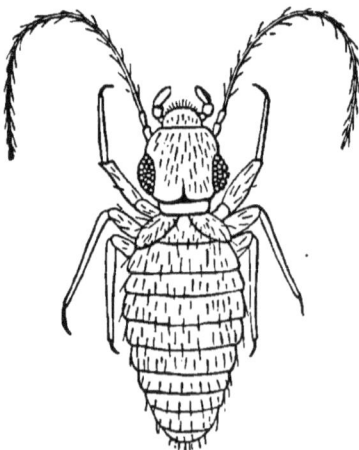

FIG. 61.— The Death-watch, CLOTHILLA PULSATORIA, greatly enlarged.

ever, be accepted as a strictly accurate one. It is a small white insect, often found in old papers, books, exposed collections of insects, etc. It occurs both in this country and in Europe, and has obtained, in the latter, the common name of the *Death-watch*, from the tapping noise which it produces resembling the ticking of a watch. To the superstitious of past ages the tick, audible only in quiet and frequently heard in the silence of a sick-room, proceeding from an invisible source, was regarded as an ill omen, predicting approaching death. A small wood-boring beetle occurring in Europe, *Anobium tesselatum* Fabr., is also sometimes known as the *death-watch*, and two or three other species of the genus have the ability of producing the same ticking sound.

Literature of the Psocidæ.

The following is some of the more easily accessible literature of the *Psocidæ* that may be consulted by those who would know more of this interesting group, and acquaint themselves with some of our species:

WESTWOOD: Introduc. Class. Ins., ii, 1840, pp. 17–20 (general notice of family).

GLOVER: in Rept. Comm. of Agricul., 1858, p. 263 (brief notice).

HAGEN: Synop. Neurop. N. A., 1861, pp. 7–14 (describes 18 species): in Proc. Ent. Soc. Phil., ii, 1863, pp. 167–8 (observations on 6 species); in Ent. Month. Mag., ii, 1865, p. 122 (Clothilla annulata); in Psyche, iii, 1881, pp. 195–6, 207–210, 219–223 (Psocina of the U. S.): id., iv, 1883, p. 25 (tarsal and antennal characters); in Stett. Ent. Zeit., 1883, pp. 285–332, pl. 2, in S. E. Z., 1882 (Monographie der Psociden).

FITCH; In Trans. N. Y. St. Agricul. Soc., xxii, 1862, pp. 668–675; 8th Rept Ins. N. Y., in 6th–9th Repts, 1865, pp. 186–193 (five new species of Smynthurus described and noticed).

WALSH: in Proc. Ent. Soc. Phila., ii, 1863, pp. 182–186 (describes 7 n. sp., and remarks on others).

HARRIS: Entomolog. Corr., 1869, pp. 327–332 (describes 7 species).

PACKARD: Guide Stud. Ins., 1869, pp. 588–590, fig. 573 (general remarks).

SCUDD.-BURG.: in Psyche, ii, 1877, pp. 49–51, 87–89 (head-structure of Atropos).

ASHMEAD: in Canad. Entomol., xi, 1879, pp. 228–9 (P. citricola, n. sp.); Orange Insects, 1880, pp. 37–38 (orange Psocus), 71–2 (habits of P. venosus).

McLACHLAN: in Entomol. Month. Mag., xix, 1883, pp. 181–185 (descriptions and generic criticisms).

AARON: in Trans. Amer. Ent. Soc., xi, 1884, pp. 37–40 (new species in Coll. of Soc.).

Achorutes nivicola (Fitch).

The Snow-Flea.

(Ord. NEUROPTERA: Fam. PODURIDÆ.)

Podura nivicola FITCH: in Amer. Quart. Journ. Agricul.-Sci., May, 1847, v, p. 284; Id., for September, 1847, vi, p. 152; Winter Insects of Eastern New York (sep. from prec.), pp. 10–11.

Podura nivicola. ASHTON: in Proc. Ent. Soc. Phil., i, 1861, p. 32.

Podura nivicola. WALSH-RILEY: in Amer. Entomol., i, 1869, p. 188 (in Wisconsin).

Achorutes nivicola PACKARD: in Fifth Ann. Rept. Peabody Acad. Sci., July, 1873, pp. 29–30.

Specimens of this winter insect were réceived from Mr. John M. Dolph, for name, with the information that they were found abundantly at Port Jervis, N. Y., on the 25th of January, both upon the surface of the snow and in pools of water made by the melting snow.

Although this is said to be a common and abundant insect, yet it would appear to be rather a local one, for, when occurring in large numbers upon the snow, it could scarcely fail of arresting attention. It certainly is not often observed, even by entomologists. I can only recall one instance in which it has fallen under my observation.

Habits.

Dr. Fitch has remarked as follows of it: "This is an abundant species in our forests in the winter and forepart of spring. At any time in the winter, whenever a few days of mild weather occur, the surface of the snow, often, over whole acres of woodland, may be found sprinkled more or less thickly with these minute fleas, looking, at first sight, as if gunpowder had there been scattered. Hollows and holes in the snow, out of which the insects are unable to throw themselves readily, are often black with the multitudes which here become imprisoned. The fine meal-like powder with which their bodies are coated enables them to float buoyantly upon the surface of the water, without becoming wet. When the snow is melting so as to produce small rivulets coursing along the tracks of the lumberman's sleigh, these snow-fleas are often observed, floating passively in its current, in such numbers as to form continuous strings; whilst the eddies and still pools gather them in such myriads as to wholly hide the element beneath them."

Notices of its Observation.

Mr. T. B. Ashton, of Washington county, N. Y., has given the following account of his observation of this insect, in a paper read before the Entomological Society of Philadelphia:

Podura nivicola Fitch. — Found on the 18th of April, weather cloudy and cold, with temperature above 50° Fahr. This insect was met with in countless numbers on and near a swampy piece of ground, through which ran a small creek. My attention was first directed to what I supposed to be soot floating down the creek, and paid no farther notice to it until I discovered the insect in large numbers in the highway, a few rods distant from the creek, and then suspecting the cause of the soot-like appearance floating on the water, I returned, and to my surprise, found countless millions of them alive and active, piled upon each other to the height of half an inch, and in spots varying from an inch or less to twelve inches in diameter, floating on the water in every eddy, for a distance of about thirty rods.

I also observed them in vast numbers, in every direction for rods around the creek. This was the only place that I met with them on that day, though I passed over, on foot, a tract of country fifteen miles in extent.

The following notices, in all probability referring to this species, are from *Field and Forest*, vol. ii, 1877, pp. 146–7.

Charles M. Nes, of York, Penn., writes to the Smithsonian Institution, that with the snow-fall of the 8th of January (about twelve inches) there appeared myriads of "spring-tails," or *Poduræ*, samples of which were inclosed, covering the surface of the snow to such an extent as entirely to discolor it. The phenomenon extended over an area of country two miles in length, and half a mile in width. In a later communication Mr. N. says: "They were in clusters, and where I gathered the specimens, I had simply to take them up by handsful; the snow was literally covered. They still exist in great quantities on fences, bushes, stones, etc., in the vicinity where they first fell.

Dr. J. G. Morris, of Baltimore, also reports a similar appearance about ten miles north of Baltimore, about the same time; and last season they were observed in numbers near Sandy Spring, Maryland.

Description.

The original description and comparisons are the following:

Black or blue-black: legs and tail dull brown.
Length, 0.08 inch.
Body black, covered with a glaucous blue-black powder but slightly adherent, and sparingly clothed with minute hairs; form cylindrical, somewhat broader towards the tail. *Antennæ* short and thick, longer than the head. *Legs* above blackish, beneath dull brown and much paler than the body. *Tail* of the same color with the venter, shortish, glabrous on its inner or anterior surface, with minute hairs on the opposite side; its fork brownish.

Though found in the same situations as the European *P. nivalis*, ours is a much darker colored species. Say's *P. bicolor* is a larger insect than the one under consideration, and differs also in size and in the color of the tail or spring. From the habits of the present species, we should infer that it might be abundant in all the snow-clad regions of the northern parts of this continent; it may, therefore, prove to be identical with *P. humicola* of Otho Fabricius (*Fauna Grönlandica*) of which we are unable to refer to any but short and very unsatisfactory descriptions, which do not coincide well with our insect (*Winter Insects of Eastern New York*, pp. 10–11).

Other closely allied species of the genus having subsequently been described, the above description would hardly suffice for its positive identification. Its more detailed description by Dr. Packard is therefore given:

Achorutes nivicola. Podura nivicola Fitch, Winter Insects, N. Y.
Antennæ four-jointed, short, thick, of nearly even width throughout; first joint as long as thick, second and third both of the same length, but second considerably thicker; fourth a little longer than third, and but slightly thinner. Feet with a large claw arising from a small joint of the usual form, with a long, slender, tenent hair arising near the claw. The anal appendages appear to be really two-jointed, and consist of two spine-like appendages, with a common base and hollow at basal half. Just below their middle is a joint, with a very slightly marked suture externally. They are borne on a broad triangular tergite (that of the

sixth segment), while low down on the side of the body are the anal valves, rounded and small compared with the tergite. Tenaculum [the catch for holding the spring] with basal joints quadrate in profile, the external edge being a little roughened. From between them project V-like a pair of short, subtriangular blades, a little longer than broad, and with four large teeth on the under side. Elater [the spring] short and broad, the two finger-shaped joints about twice as long as thick; second joint very minute, consisting of a thin lobe rounded at tip and a little uneven on lower edge, and with a prominent spine at base; two hairs on inside of basal joints. Dark lead color, above and beneath. Body long and slender, abdomen rather suddenly contracting just before the three terminal segments.

Length, 0.08–0.10 inch.

Hundreds of specimens seen with no appreciable variations. Very young, white.

On snow in March; under bark of trees. Salem, May 28 and June 6. (Packard: in 5th Rept. Peab. Acad. Sci.)

Associated Species.

The following named features will serve to separate the above from the other described species:

Achorutes boletivorus Packard, is a much smaller species, being only about one-half the size (0.05–0.07 inch) with a thicker body, smaller head proportionately, and shorter antennæ. It differs materially in color, being pale gray above with a slight greenish tinge, lined with white beneath, and with gray specks on the sides.

Occurring September 10, abundantly between lamellæ of Boleti and Agaricus, and under horse manure in August.

Achorutes pratorum Packard. Resembling *A. nivicola*, but smaller (0.04–0.06 inch), of a considerably lighter lead color, paler beneath, and with a lateral row of dark irregular spots, one to each segment. The hairs upon the body are more numerous. Its spring (elater) is much longer and more slender, the second joint long, slender, cylindrical, distinctly separated from the base by a suture; the third and terminal joints much longer and more slender. It occurs on the surface of pools, and in open grounds after rains, and is quite active in its habits. July and September.

Achorutes marmoratus Packard. Approaches the preceding in form of body and of the spring. It is of a pale gray color, marbled with large lilac-gray patches above; beneath paler. Length 0.05 inch.

Achorutes Texensis Packard. A small species (0.04–0.05 inch), occurring in Texas, of the same color with *A. nivicola*, resembling *A. pratorum* in structure, but with its spring twice as large as in that species.

The Family of Poduridæ.

The *Poduridæ*, from their degraded structure, form one of the lower classes of insects. By most writers, they have been classed with, and at the end of, the Neuroptera, but by some systematists they have been united with the *Lepismatidæ* (bristle-tails) into a separate order, known as Thysanura. They have been so referred by Dr. Packard in the last edition (8th, of 1883) of his *Guide to the Study of Insects.*

They are small species, varying from 0.04 to 0.16 of an inch in length. The largest species, *Orchesella flavo-picta* Packard, occurs in Albany, under bark of stumps.

They are commonly known as "Spring-tails" from the flexible anal stylet which is bent beneath the body, serving as a spring, by means of

which they are enabled to make extraordinary leaps when compared with their diminutive size. They are commonly found

FIG. 62.—ACHORUTES PURPURASCENS —enlarged to 18 diameters (After Murray).

beneath bark, stones, sticks, boards, in crevices of wood, in cellars, in garden hotbeds, on snow, pools of water, damp earth, manure, Boleti and Agarics, and in various other similar locations.

As illustrative of the genus to which the snow-flea belongs, an European species, *Achorutes purpurascens* (Lubbock), is copied (Fig. 62), from Murray's *Economic Entomology.* In Fig. 63, another view, copied from *Science Gossip*, of the same insect is presented showing its under surface, and the short forked spring characterizing the different species of this genus.

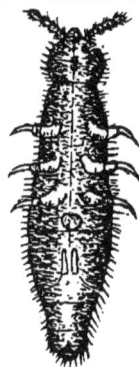

FIG. 63. — ACHO- RUTES PURPURAS- CENS, showing un- der side.

A different type of the *Poduridæ* is shown in Fig. 64, in *Smynthurus hortensis* Fitch, in which the short and nearly spherical form approaches the spiders. It occurs abundantly in May and June,

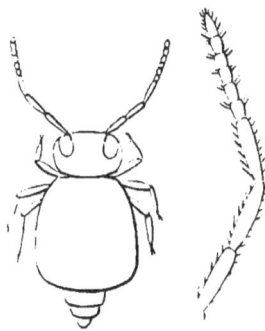

FIG. 64.—The Garden-flea, SMYNTHU- RUS HORTENSIS Fitch, greatly enlarged, with antennæ still more enlarged (after Fitch).

in gardens, in the State of New York, upon the leaves of young cabbages, turnips, cucumbers, and many other plants. It is believed to be injurious to the vegetation upon which it is found, through its continuing the attack made by the flea-beetles and other insects of similar habits, and enlarging the wounds or perforations made by them, by feeding upon the soft matter formed by the evaporation of the exuding juices. It is an active insect, and quickly skips from the leaf, if disturbed, to the ground. For an interesting description of the peculiar apparatus — the forked spring and catch — by means of which its leaps are made, see Fitch's 8th Report (6th–9th Reports, p. 188).

Lipura fimetaria (Linn.).

(Ord. NEUROPTERA: Fam. PODURIDÆ).

Podura fimetaria LINN.: Syst. Nat., 12th edit., 1766, p. 1014.
Lipura fimetaria PACK.: in Amer. Naturalist, v. 1871, p. 106, f. 38; in 5th Rept.
   Peab. Acad. Sci., 1873, p. 29.
Lipura fimetaria. MURRAY: Econom. Entomol.—Aptera, p. 412, f. 21.
Lipura fimetaria. LINTNER: in Count. Gent., xliv. 1879, p. 327; id., xlv, 1880,
   p. 103.

This little species of the *Poduridæ*, frequently found in damp localities, but scarcely regarded as of economic importance, has in two instances been brought to my notice under circumstances that made it of
more than usual interest — in one instance abounding in a cistern, and
in the other in a well used for household purposes.

With the first, the following communication was sent:

Its Occurrence in a Cistern.

Please find inclosed some strange insects that I found in my cistern
about three weeks since [in January]. The cistern is a large one, with a
small receiving cistern near it, from which the water is filtered into the
large one. During the last severe weather the pump froze and broke.
I then ordered water to be drawn with pole and bucket. The first
bucketful drawn up appeared to have hundreds of these little white insects in it, and they continued to come up with every bucketful. These
I have had in a glass goblet in a warm room for the last three weeks.
The cistern has been built a little over two years, and has a close-fitting
iron cover on it. There is no wood in or around it. Where did these
insects come from?

The insects were alive when received, and although they had been
confined for two days, at least, in a homœopathic vial, they were found
quite active upon the surface of the water. Their tenacity of life is
somewhat remarkable, for it is recorded of this species that it has been
kept in confinement from September until the following June.

In reply to the communication, assurance was given of their harmless
character in the location found, as they could only act as scavengers,
serving to remove impurities. In answer to the question where they
came from, it was stated that the species occurred under damp sticks or
wet pieces of wood. A favorable locality for it would be under the
roof shingles projecting over a wooden gutter, through which, washed
off by the rains, they could be readily conveyed into a cistern, and,
from their fecundity and rapid propagation, soon become greatly multiplied. As they live upon the surface of the water, their presence would

not be detected so long as a pump was used unless the water had been reduced to the level of the bottom of the pipe. A bucket dipping from the surface would at once bring them to notice.

Occurrence in a Well.

Another communication contained the following statement:

I inclose to you by mail in a bottle, four or five small insects which are drawn up in the bucket of my well quite frequently. Can you inform me what they are, and in what way they can be most easily exterminated? The well is dug in a sandy clay, and is made of two and a half foot cement pipe, the joints being cemented, and I fully supposed it was worm, insect and vermin proof. At times hundreds are drawn up in a day. They seem to live only on the surface of the water.

As to their occurrence in the well, it was suggested in reply, that if the well was an open one, the insects would originally have been attracted by the cool, damp sides of its walls, and thence easily have found their way to the water. It was no doubt somewhat disagreeable to see living forms in association with water which we drink, but beyond this, the presence of the *Poduridæ* in the well was not objectionable. As in the case of those occurring in cisterns, they would undoubtedly serve an excellent purpose in feeding upon and removing many of the impurities that, without them, would accumulate upon the surface, to the detriment of the healthfulness of the water. In the event, however, of their becoming annoyingly abundant, it was thought that they could be diminished by dropping a moderate quantity of finely powdered lime into the well.

Its Appearance and Habits.

We find among our writers no detailed description of this species, nor good figure, and we therefore copy a rather crude one from Murray's *Economic Entomology*, to aid in its identification. Fig. 65 gives a dorsal view of the insect. Although contained in the Tribe Collembola, which consists largely of the " spring-tails," this species is without the jumping apparatus of the one previously noticed. Dr. Packard, in his *Synopsis of the Thysanura of Essex Co., Mass.*, places it in the subfamily of *Lipurinæ*, and characterizes it briefly as being white, naked, with a few scattered hairs, and of the length of 0.06 of an inch. It is rather blunt at the tip, and is without the hooks at the end of the abdomen that occur in an associate species, *Lipura ambulans* (Linn.).

FIG. 65.—LIPURA FIMETARIA (Linn.). enlarged to ten diameters (After Murray).

27

Its habits appear to be quite varied, for beside living upon the surface of water after the manner of *Anurida maritima* (Linn.), *Podura aquatica* Linn., and others — according to Murray, it is found upon damp earth, in Europe, throughout the year, often engaged in browsing upon carrots, potatoes, or other roots.

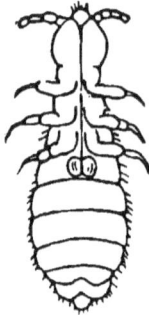

FIG. 66.—ANURIDA MARITIMA, enlarged 25 diameters.

Fig. 66, copied from an article upon some English Thysanura in *Science Gossip* for 1873, is of *Anurida maritima* — a small, dark blue velvety Podura, inhabiting the wet sea-weed, the loose shales between tides, and the rock-pools of the English, Irish and French coasts; also upon the eastern sea-coast of the United States, at Salem, Mass., at Nantucket, and on the New Jersey coast, floating on seaweed, or hidden under stones between tide marks (Packard). The figure gives a ventral view of the insect, and shows the absence of the abdominal spring — a distinguishing feature which separates the genera of *Anurida*, *Lipura*, and *Anura* from the other *Poduridæ*.

APPENDIX.

(A.)

ENTOMOLOGICAL CONTRIBUTIONS.

The two following papers are republished in this place, for convenience of reference, and as being a portion of the work of the Entomologist during the period for which his report is made.

[*From Psyche, Novem.-Decem.*, 1883, iv, *pp.* 103-106.]

A NEW SEXUAL CHARACTER IN THE PUPÆ OF SOME LEPIDOPTERA.

(Read before the American Association for the Advancement of Science at its Montreal
meeting, August, 1882.)

The sexual characters of insects have always been an interesting study
to the entomological student, the more so as they are the less apparent,
and discoverable, if to be found at all, only as the result of close observation and comparison. In the larger proportion of insects, in the perfect
stage, they are so marked as to leave no doubt of the sex when the male
and female are compared. Thus in the Hymenoptera, we have the ovipositor in its varied forms, often quite conspicuous. In the Lepidoptera;
among the *Heterocera*, there are usually the more fully developed antennæ
of the male, and the broader, conical and more capacious abdomen of the
female — features attaining their maximum development in the family of
Bombycidæ. In the Diptera, there are the larger and more approximate
eyes in the male, and conspicuous structural differences in the antennæ
and suctorial apparatus in some of the families. In the Coleoptera, there
are often, in the male, stouter legs, broader tarsi, greatly elongated mandibles and other horn-like caputal and thoracic processes. In the Hemiptera, the vocal organs in the *Cicadidæ*, the ovipositor in several of the
families, and the great sexual differences in size and in the presence or
absence of wings are prominent features. In the Orthoptera, there are
the stridulating wing-nerves, the extended ovipositor, and a genital armature greatly varied in its adaptation to greatly differing habits. And in
the Neuroptera, distinctive male characters are found in clasping organs,
in differences in color and in size, the long mandibles of *Corydalus*, the
abnormal location of the intromittent organ in *Libellulidæ*, and in the
elongated and forcipated genitalia of *Panorpa*.

In addition to such primary features as above noted, there are numerous secondary ones, which do not appear to be so dependent upon sex,

and many of which seem almost to serve no higher purpose than that of ornamentation. Yet it is reasonable to believe that most of these differences have their use in the economy of nature, and that they aid in the continuance of the species.

Among such minor antigenetical features may be mentioned, in the Lepidoptera, the usually more angulated wings of the male; the simple frenulum of the most of the male *Heterocera* in contrast with the compound one in the female; the hairy anterior legs of *Grapta* and *Vanessa* in the *Nymphalidæ;* the long hairs between the costal and subcostal nervures, above the cell of the hind wings of *Argynnis*, appearing when displayed in the cabinet, like a long fringe to the inner margin of the front wings; the incrassated, black, scale-patch upon the middle of vein 2 (the 1st median nervule) of the secondaries of *Danais;* the ovoid discal spot on the front wings of many of the *Theclinæ:* in the *Hesperidæ*, the reflexed costal margin in most of the *Nisoniades, Eudamus* and *Pyrgus*, and the tibial epiphysis* of the anterior legs in all but one of our genera; the transverse discoidal stigma on the primaries of most of the species of *Pamphila;* the beautiful and peculiar microscopic (often concealed) scales, or androconia, of many of the butterflies; the usually concealed pair of extensile anal appendages found by Fritz Müller and others in certain *Glaucopidæ, Bombycidæ, Noctuidæ* and in a *Danais,†* — each of these several characters indicating the male sex. Features equally interesting, and alike serving no purpose so far as known, might be mentioned in each of the orders of insects.

In the earlier stages of insects (egg and larval), sexual features, as would naturally be expected, are less numerous and less conspicuous. They rarely occur in the first stage — that of the egg, or more properly, they have not, in many instances, been recognized by us.‡

It was for a long time believed that in the larva of one of our *Sphingidæ* not unfrequently met with — *Thyreus Abbotii* — the sex was so clearly indicated by difference in color and pattern that it could be told at a glance. Of the two greatly differing forms, the one marked with a series of large yellow-green patches on the dorsum extending half-way down the sides, and with another row of smaller subtriangular similarly colored spots resting on the prolegs, was described by Clemens as the male; the female being reddish brown throughout, with a dark brown subdorsal stripe and numerous short broken striæ.§ This sexual determination of

* Guenée: Hist. Nat. Ins., 1852 — Lepid., v. — Noct., i, p. xxxv.—Speyer: in Canad. Entomol., 1878, v. 10, p. 124. Edwards' Catal. Lep. Amer., 1877, p. 64.

† Fritz Müller: Nature, 11 June, 1874, v. 10, p. 102 (Psyche, Mar.-Apr. [9 July], 1877, v. 2, p. 24). Morrison: Psyche [9], Oct. 1874, v. 1, pp. 21–22. Siewers: Canadian Entomologist, Mch. 1879, v. 11, pp. 47-48, fig. 12. Stretch: Papilio, Feb., 1883, v. 3, pp. 41–42, fig.

‡ In *Phylloxera*, the eggs which are to produce males and females may be known by their difference in size. See Riley's Annual Reports of the State Entomologist of Missouri: 6th, p. 41; 7th, pp. 92–98; 8th, p. 158.

§ Two colored figures of the larvæ in my possession, made by Dr. Clemens, show the sexes the reverse of this — the green-spotted one marked as ♀, being much the larger of the two.

Clemens was accepted by me in my paper upon the larvæ and pupæ of this species in the 26th Report of the New York State Museum of Natural History, pp. 114–116, and has also been followed by other writers. That the two forms are indicative of sex, has since been denied,* and it is to be presumed that the denials are based upon results obtained in rearing them to their perfect form. The green-spotted larva may, therefore, be accepted as a dimorphic form, comparatively rare in my own collections and in the examples that have come under my observation.

The young collector of insects learns very early the simple method of determining the sexes of his Luna, Polyphemus, Promethea, and Cecropia pupæ, and of many other bombycid pupæ, by observation of the comparative breadth of their antennal cases.

A means by which the sex in the pupæ of the *Sphingidæ* may be infallibly named, was pointed out by me in the *Proceedings of the Entomological Society of Philadelphia*, 1864, v. 3, p. 654. I have since found the same characters applicable to the *Noctuidæ* and to other *Heterocera*.

Prof. C. V. Riley, in the *Transactions of the Academy of Science of St. Louis*, 1873, vol. 3, pp. 128–129, and in the *6th Annual Report of the State Entomologist of Missouri*, (or 1873, 1874, pp. 131–132, has described and figured sexual differences in the pupæ of *Pronuba yuccasella*, consisting mainly, in the greater length of the " dorsal projections " on the several segments of the male, in the length of the last two segments as compared with those of the female (its shorter 11th and longer 12th), and in its less rounded apex. He says: "Sexual distinctions are very rarely observable in chrysalids ; but after I had learned to distinguish between them, I could readily separate the sexes in this case, and my judgment was confirmed upon the issuing of the moths."

A few years ago I discovered an interesting feature in the armature of the species of *Cossus*, by which the sex may at once be determined. I have hitherto withheld its publication, until I had studied others of our spined pupæ and could illustrate this feature by proper figures ; but the opportunity for this has not been found, and I accordingly defer no longer calling attention to it, that the observations of others in possession of more abundant material may supplement the few that have been made by me.

It is known to lepidopterists that most of the pupæ of the species of moths which in their larval stage live in the interior of stems of plants and trunks of trees (endophytes), are armed upon their abdominal segments with transverse rows of teeth or spines, by the aid of which, when they are in readiness for their final transformation, they gradually work their way through the outer packing of their gallery and the bark, project their anterior segments to at least one-third the entire pupal length through the opening, and hold themselves securely during the eclosion of the moth.

* Whitney : Canadian Entomologist, April, 1876, v. 8, pp. 75–76. Grote: *id.*, May, 1876, p. 100.

This useful armature in the *Cossinæ*, and in such of the *Ægeriidæ* as I have had the opportunity of examining, consists of two rows of spines upon most of the abdominal segments, dividing them, when seen in extension, in three nearly equal parts. In *Cossus robiniæ*, the species of the . *Cossinæ* with which we are probably the most familiar, these rows occur on the fifth (the first stigmatal segment posterior to the wing-cases) and the following segments.

In *Cossus querciperda* alone of the species known to me, they commence in a single row of minute dentations on the fourth segment. The principal features of this armature are the following : It is always the stronger in the male sex — conspicuously so in *C. robiniæ*, but less so in *C. Centerensis:* the teeth increase in size from the fifth to the tenth segment : the anterior row is always the stronger in each sex ; upon the fifth and sixth segments, it does not, in its lateral extension, reach below the stigma,* while upon the following segments it passes in front of the stigma and quite a distance beneath it ; the posterior row is discontinued before reaching the line of the stigmata ; the teeth show irregularity in form and size, particularly those of the posterior row.

The sexual distinction above referred to, presented in this armature, is this : in the male pupæ two rows of teeth occur on segments five to ten inclusive ; in the female, two rows on five to nine inclusive. In other words, *the male pupa shows* TWO *rows of teeth on segment ten, where the female shows but* ONE [as illustrated in Fig. 67, of a pupa-case of *C. Centerensis*, from which the moth had emerged]. In each sex, the eleventh and twelfth have but a single row. Disregarding, as I think we

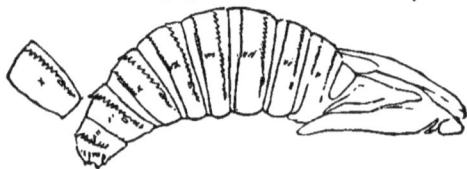

Fig. 67.—Pupa-case of Cossus Centerensis, male (the moth emerged), with tenth segment of a female pupa-case—enlarged.

should in ordinary usage, the subdivision of what is usually known as the terminal segment, into demi-segments, or a segment and a subsegment, and that still further refinement which would make of the extreme portion an additional segment with full numerical designation, then it will serve to prevent misapprehension of the particular section showing the sexual feature, if we indicate it as the *antepenultimate segment.* It would be the eleventh, if we commence enumeration, as some of our entomologists do, with the head, but the tenth, if, as seems to me more proper, we begin with the first thoracic ring.

Besides the *Cossinæ*, this same sexual feature occurs in the *Ægeriidæ.* I am not able to say if it extends throughout the entire family. At the time of this present writing, I have at my command only the pupæ of *Ægeria exitiosa* and *A. tipuliformis*, and it exists in each. It probably occurs in the pupæ of *Zeuzera* (one North American species described), in which the two rows of teeth are found on several of the segments, and per-

* In *C. Centerensis* it reaches below the stigma on the sixth segment.

haps also in *Hepialus*, the pupæ of which (unknown to me) are characterized as very similar to those of *Cossus*.

Another interesting fact connected with the armature of *Cossus* is that
the form, size and position of the teeth vary to so great an extent in the
different species, and show such distinctive characters, as to afford excellent specific features.[*] I would not hesitate to pronounce upon specific
identity, upon an examination and comparison of the pupal armature
alone.

[*From Psyche, May-June, 1883, iv, pp. 48-51.*]

ON AN EGG-PARASITE OF THE CURRANT SAW-FLY (NEMA-
TUS VENTRICOSUS).

(Read before the American Association for the Advancement of Science, at its Montreal meeting,
29th August, 1882.)

Dr. Asa Fitch, in his *12th Annual Report on the Insects of New York*
for the year 1867 (*Trans. N. Y. State Agric. Soc.* for 1867, 1868, v. 27), pp.
931-932, made the following reference to this insect :

As none of the foreign accounts which we have seen allude to any
parasitic enemy of this currant saw-fly, it seemed quite improbable that it
would in this country meet with any such enemy, to lighten from us the
task of combatting it and diminishing its devastations. But our valued
friend, J. A. Lintner, of Schoharie, greets us with the glad tidings that he
has discovered we have such a foe to this formidable scourge. An egg
parasite of this saw-fly inhabits our State, an exceedingly minute hymen-
opterous insect, which inserts its eggs into those of the saw-fly, that its
young may subsist upon and consume the contents of those eggs. This
diminutive little fly has probably existed hitherto upon the eggs of some
one of our American saw-flies similar in size to those of the currant saw-
fly ; and it has now discovered that the eggs of this newly-arrived foreigner
are equally well adapted to its wants. And so multiplied has this little
friend of the gardener become, that in Utica, Mr. Lintner finds that among
fifty eggs of a saw-fly upon a currant leaf, there will not be more than
four or five that will hatch currant worms, all the rest being occupied by
the little maggot, the young of this parasite. At Schoharie, also, where
the saw-fly has arrived more recently than at Utica, he finds this parasite
is now beginning to appear. Everywhere this little creature is no doubt
following upon the tracks of the saw-fly, and within a very few years
after the one arrives in any place the other will be there also, and will
speedily become so multiplied as to quell and extinguish it. This is a
most important discovery, and renders it quite probable that in this country this currant worm can never be but a temporary evil. Whenever cir-

---

[*] For comparison with other species of the *Cossina* it may be stated that an example of
C. Centrensis, ♂, has thirty-eight teeth in the anterior row of the tenth segment, and
twenty teeth in the posterior row — the latter, in their entire range, occupying a transverse
space equal to that of nine teeth of the anterior row. The teeth are black, shining, irregular in size, and are slightly bent upward over their base ; their length and the distance
between their tips exceeds their basal width.

cumstances favor it and enable it to multiply and become numerous in any section of our country, this little enemy, its mortal foe, will speedily be there to subdue and stamp it down. Thus nicely are the works of nature balanced, and no creature is permitted to usurp a place in her domain which does not belong to it.

The specimens of the parasite obtained by me at the time referred to' in the above notice, were submitted to a friend who had made study of the group to which they belong, who believed them to be an undescribed species, and was only able to give them a doubtful generic reference. They were subsequently destroyed, and from that time until the present year (an interval of fourteen years), although I have continued to search for them I have been unable to obtain the species.

Its rediscovery by me the present year, and the determination of the species, lend additional interest to the notes upon it that I made at its first observation, at Utica, N. Y., in June, 1866, and I therefore transcribe them from my note-book :

I had collected a number of currant leaves upon which the currant saw-fly had deposited eggs, and was counting the eggs upon each to obtain the average number per leaf, when I noticed an occasional brown egg among them, appearing somewhat abnormal in shape. On placing them under a lens a resemblance to a pupal form was detected. I at once suspected the presence of the parasite for which we had been hoping. Although there seemed to be but the merest chance of discovering at large an insect so minute as this must necessarily be, I instituted a careful search of the currant bushes in the garden, and in a short time had the great gratification of seeing a minute speck moving among the eggs, which under my lens revealed a form which left scarce room for doubt of its parasitic character. During the day I detected several more of the kind upon the leaves containing egg-deposits, affording strong evidence of their relationship. A few days thereafter (perhaps a week), in a small phial in which I had placed some eggs that I suspected to have been parasitized, I had the delight of seeing several of the familiar forms of my currant-leaf acquaintances, and the ruptured pupa cases from which they had evidently escaped.

The following year (1867) there was a marked diminution in the number of currant-worms observed, and a corresponding increase in parasitized eggs. Many of the leaves had not been visited by the parasite, but of those that gave evidence of such visit, the work of destruction was almost complete, for of several leaves bearing each from thirty to forty eggs, all but five or six were transformed into parasitic pupæ.

In June, 1868, I was able to make, at Schoharie, N. Y., the following observations upon the oviposition of the parasite within the eggs of the currant saw-fly :

In a small phial in which had been placed some parasitized eggs of the saw-fly, a male and female parasite had emerged. That I might observe their actions I introduced a piece of currant-leaf having upon it some eggs which I had just seen deposited. No evidence was given that the

female was aware of the presence of the eggs, but after several minutes
traveling around the glass, she moved upon the leaf, and in passing over
and beneath it, seemed to meet with them accidentally. She paused, and
then began a careful inspection, walking over them several times, and
constantly palpating them with her antennæ. Then satisfied with her
examination, she attached herself to one of the eggs, appressed the tip of
her abdomen to it, and remained in this position motionless for the space
of two and a half minutes, during which time an egg, doubtless, was inserted, although the pocket lens with which the observation was made
did not disclose the fact. The motion of her antennæ then recommenced
and I expected to see the operation just witnessed repeated upon another
egg; but, to my surprise, she merely changed position — again applied
the tip of her abdomen to a different part of the same egg, and remained
at rest for about the same space of time as before. Three times I witnessed this performance, and it is, therefore, probable that three parasitic
eggs were placed within the one of the currant-fly. Unfortunately an in-
terruption prevented me from noticing if the remaining currant-fly eggs
were similarly parasitized, and the number of eggs introduced in each ;
and much to my regret, the eggs were accidentally destroyed before my
observations could be made upon their transmutation into parasitic pupæ.
The pupa cases are dark brown, disclosing some of the outlines of the
contained pupæ, somewhat flattened, broader than the original egg, but
of about its length. The insect is apparently one of the *Chalcididæ*, having a broad head, long and elbowed antennæ, ovoid anterior wings, nearly
veinless, beautifully iridescent, delicately fringed and haired ; the posterior wings are almost linear; the abdomen is short, not reaching the
tips of the wings.

This year (1868) is probably the first appearance of the parasite at Schoharie, as I could only discover about a dozen individuals. Its progress
seems to be from west to east, corresponding with that of the currant-
worm.

The rediscovery of the parasite the present year (1882) was made in my
garden at Albany, upon a solitary currant bush growing there. The parasitized eggs were inclosed in a bottle, and in a few days the insects emerged.
That I might multiply and aid in the distribution of an insect which had
already shown its capability for usefulness, I visited another garden in the
city to obtain eggs of the currant-fly for parasitization by my confined
individuals. To my surprise, the parasite was here found in strong force,
for in the examination of a long row of currant bushes containing many
eggs, I could not find a single egg-bearing leaf which had not been visited
and the destruction of the eggs insured. A large number of leaves were
collected, each bearing perhaps from forty to fifty parasitized eggs. Re-
serving a few of these for study and for propagation, the remainder were
made up in small parcels of about a half-dozen each, and mailed to entomological friends in various parts of the United States and Canada,* with

[* Canadian Entomologist, xiv, August, 1882, p. 147.]

the request that they be pinned upon currant bushes among the leaves where the currant-fly eggs were to be found. The introduction of parasites in this manner into localities where they had not previously occurred, has been shown to be practicable ;* and in consideration of the great importance of parasitic aid in the destruction of our insect pests, I sincerely hope that my efforts to distribute this very efficient parasite may prove, from observations to be made hereafter, to have been successful.

Examples of the insect were sent by me to Mr. L. O. Howard, of the Department of Agriculture at Washington — a gentleman who has made special study of the family to which it pertains, viz., the *Chalcididæ*. He informs me that there is no doubt of its being the species described and named by Prof. C. V. Riley, in 1879 (Can. Entom., Sept. 1879, v. 11, pp. 161–162) as *Trichogramma pretiosa*, examples of which had been reared, at Washington, from eggs of the cotton-worm moth, *Aletia argillacea* Hübn., collected in Alabama. The description is reproduced, with additional information, in Prof. J. H. Comstock's Report upon Cotton Insects (Washington, 1879), p. 193. It has since been extensively reared from eggs of the same moth collected in Florida, by Mr. H. G. Hubbard. It has also been bred at the U. S. Department of Agriculture from eggs of an unknown Noctuid moth occurring on orange trees, and from *Aleyrodes*.

Prof. Riley, from some structural features, thought that it might be necessary to establish a new genus for this species and one or two closely allied ones, but Mr. Howard finds it to be a true *Trichogramma*, as at first referred.

Another species of the genus *T. minuta* Riley,† [shown in Fig. 68, and

FIG 68.—TRICHOGRAMMA MINUTA Riley: *a*, the fly in its natural position; *b*, a front wing; *c*, a hind wing; *d*, one of the legs; *e*, an antenna—all much enlarged.

hardly to be distinguished in appearance from *T. pretiosa*], has been reared from the eggs of one of our common butterflies, of extensive distribution, *Limenitis disippus*. Parasitized examples of these eggs have given from four to six specimens of the minute creature, which, notwithstanding its specific name of *minuta*, exceeds in size the microscopic *T. pretiosa*, the latter being about 0.25 mm. in length.

In connection with the above notice of the egg-parasite of the currant-fly, it may be of interest to offer the following note of the oviposition of the currant-fly as observed by me, as its method has not to my knowledge been previously published.

June 7, 1868. *Nematus ventricosus* was seen to deposit thirty eggs upon a single currant-leaf within one hour. In the act of ovipositing, it curved the tip of its abdomen downward and forward, directing its ovipositor toward its head, in which position the end of the egg is seen to protrude

*Le Baron: *Third Annual Report on the Insects of Illinois*, 1873, pp. 200–202.
†*Third Annual Report on the Insects of Missouri*, 1871, p. 158, fig. 72.

and attach itself to the leaf-nervure, when the ovipositor is withdrawn, and the egg left in position. Moving backward a very little, another egg is similarly deposited, and in like manner the operation is continued, until the leaf has its assigned quota, or the supply of eggs is exhausted. The eggs produced their larvæ on June 14th.

(B.)

MISCELLANEOUS PUBLICATIONS OF THE ENTOMOLOGIST.

The following list comprises publications during the years 1882 and 1883, mainly in agricultural journals, most of which were in reply to inquiries made in relation to insects of economic importance:

Insects on Sweet Potato Vines. (Country Gentleman, for Feb. 23, 1882, xlvii, p. 149, c. 2–3 — 12 cm.)

The larvæ infesting sweet potato vines in St. Louis, Mo., are probably those of *Coptocycla aurichalchea*. Remarks upon the species and means of destruction, viz., application of Paris green and London purple.

Entomological — The Anatomy of the Mouth Parts and the Sucking Apparatus of some Diptera. (Country Gentleman, for February 23, 1882, xlvii, p. 151, c. 2–3 — 13 cm.)

The paper was prepared by Mr. George Dimmock, of Cambridge, Mass., as a dissertation for obtaining the philosophical doctorate at Leipsig University. In it the mouth-parts of *Culex, Bombylius, Eristalis,* and *Musca* are ably treated, of, and fully illustrated in four excellent plates. A résumé of previous publications on the subject is given.

Millions of Grasshoppers in Midwinter. (Albany Evening Journal, for February 25, 1882.)

Grasshoppers observed upon the surface of snow in Westchester county, New York, prove to be *Tragocephala viridifasciata*, "the green-striped locust" of Dr. Harris. This species has on different occasions appeared during warm days in winter, but has never been recorded as very injurious. Their early appearance at this time will cause the death of large numbers, and lessen the brood of the summer months.

A Winter Grasshopper — Tragocephala viridifasciata. (Country Gentleman, for March 9, 1882, xlvii, p. 189, c. 2 — 16 cm.)

Identification of the species, occurring in Genesee county, N. Y., in February — its history — is not often injurious — will probably be killed by frosts before its food appears.

The Hickory-Borer — Cyllene pictus (*Drury*). (Country Gentleman, for March 9, 1882, xlvii, p. 189, c. 2–3 — 8 cm.)

The species was for a long time confounded with the locust-borer, C.

robiniœ, but is separable by its longer and stouter antennæ and body behind tapering to a blunt point. The specimens received are in the pupal stage, showing in transparency, the yellow markings of the wing-covers. The larva is not as injurious to hickories as the locust-borer to locusts.

Apple-Leaf Bucculatrix. (Country Gentleman, for March 16, 1882, xlvii, p. 207, c. 1 — 5 cm.)

Small white-ribbed cocoons upon apple-tree bark, sent from Bergen county, N. J., are identified as those of *Bucculatrix pomifoliella* Clemens. The cocoons show a parasitic attack.

Insects that Injure Trees. (Country Gentleman, for April 20, 1882, xlvii, p. 313, c. 1-2 — 46 cm.)

Notice of Dr. Packard's "Insects Injurious to Forest and Shade Trees," being Bulletin No. 7 of the U. S. Entomological Commission. The volume presents a summary of what is known of insect injuries to our more useful trees. The notice remarks upon the importance of the preservation of our forests and attention given in Europe to the subject, and refers to the number of species of insects attacking trees in Europe and in the United States; it complains of the inferior character of some of the illustrations ; mentions a revised edition contemplated.

The White Grub — Lachnosterna fusca (*Frohl.*). (Country Gentleman, for April 27, 1882, xlvii, p. 333, c. 2-3 — 34 cm.)

Belief expressed that the grubs destroying the roots of grass, will not injure potatoes on the same ground the following year. Methods given for their destruction, as shaking the beetles from the trees on which they congregate, rooting out the grubs by hogs, plowing up and exposing to birds, application of salt and other materials to render the food unpalatable to the grubs.

Mites in Timothy Fields. (Country Gentleman, for May 18, 1882, xlvii, p. 395, c. 1-2 — 19 cm.)

The little attention paid to the study of the Acarina in this country : reference to a few common species. This species pronounced by Dr. Hagen, of Cambridge, to be *Trombidium bicolor* Hermann, or very near to it. It is black with red legs — was observed April 28th at Concordville, Pa., on no other grass but timothy.

The Spring Canker-Worm — Anisopteryx vernata. (Country Gentleman, for May 18, 1882, xlvii, p. 393, c. 1-3 — 67 cm.)

Its distribution is from Maine to Texas, but it is usually quite local ; efficiency of birds in destroying it ; ascent of the tree trunks by the wingless moths should be prevented, may be done by bands of tarred cloth ; Dr. Le Baron's band of rope with tin nailed upon it, and how it operates; another trap (illustrated) is a band of tin suspended from a cloth bound to the trunk; when caterpillars are on the tree, beat them down on straw and burn them, or spray the leaves with Paris green water by a force-pump. Protection from the flat-headed apple-tree borer.

Leaf-mining Anthomyiidæ. (Canadian Entomologist, for May, 1882, xiv, pp. 96–7.) (Thirteenth Ann. Rept. of the Entomological Society of Ontario, for the year 1882. 1883, p. 29.)

Discovery of the first American species of leaf-mining *Anthomyiidæ*, mining beet-leaves at Middleburgh, N. Y., viz., *Chortophila floccosa* Macq., and two new species, to be described in the forthcoming Report of the N. Y. State Entomologist.

The Grain Aphis — Siphonophora avenæ (*Fabr.*). (Country Gentleman, for June 22, 1882, xlvii, p. 493, c. 2–3 — 22 cm.)

Infesting wheat in Virginia; Dr. Fitch's notice of it in his 6th Report; its attack on the heads continued until the kernels harden; many of the examples are parasitized; *Aphis granaria* a synonym; attacks also oats, rye and barley.

The Apple-Tree Case-Bearer. (Country Gentleman, for July 6, 1882, xlvii, p. 533, c. 1–2 — 28 cm.)

The caterpillar, bearing its peculiar case, is sent from South Byron, N. Y.; natural history of the species, known as *Coleophora malivorella* Riley, given, with references to full notices of it; spraying with Paris green and London purple in early spring a good remedy for it. Directions for mailing injurious insects — should not be sent in paper boxes, permitting escape and propagation in new localities.

The Spring Canker-Worm — Anisopteryx vernata *Peck*. (Country Gentleman, for July 6, 1882, xlvii, p. 533, c. 2–3 — 26 cm.)

Abundance of the caterpillars in New Canaan, Ct., confined to two orchards. For means of destruction reference is made to notice in the C. G., of May 18. The pupæ, buried three or four inches deep beneath the tree, may be killed by breaking up the ground, or by turning swine in the orchards to root them up.

The Rose-Bug. (Country Gentleman, for July 6, 1882, xlvii, p. 534, c. 3 — 10 cm.)

Leaves stripped from cherry trees in Scarsdale, N. Y., by *Macrodactylus subspinosus*. The inquirer of name and habits is referred to C. G., of June 26. Sprinkling foliage with tansy water has been said to prevent its depredations. Paris green sprinkling would destroy it; shaking it from trees on sheets recommended.

The Seventeen-year Locust. The Ontario County Times, xxviii, for July 12, 1882, p. 3, c. 5 — 58 cm.)

Gives the seventeen-year and thirteen-year periodicity of Cicadas

29

and the years of their appearance during the present half-century; broods in the State of New York ; injuries of the present brood upon the shore of Canandaigua lake, and preventives of future injuries.

A New Household Pest — Attagenus megatoma. (Country Gentleman, for July 20, 1882, xlvii, p. 567, c. 2–3 — 38 cm.)

Description of the beetle and of the larva; the larvæ occur beneath carpets upon which they feed ; abundance of the beetle in Washington Park, Albany, on spiræas with *A. scrophulariæ ;* are often found on windows within our houses ; probably breeds also in hair-cloth furniture; does it injure cotton fabrics ? benzine and kerosene for killing it ; tympans and roofing-paper as preventives.

The Hessian Fly in Ohio. (Country Gentleman, for July 20, 1882, xlvii, p. 567, c. 3 — 14 cm.)

The wheat attacked by the second brood of the insect, of which the eggs were laid in May. The flies would emerge in July or August to attack the winter wheat ; may possibly be a third brood. Protection of its parasites recommended; burning stubble would destroy the parasites. Wheat broken down by the attack should be cut low and reaped.

A Bark Beetle. (Country Gentleman, for August 3, 1882, xlvii, p. 605, c. 2–3 — 18 cm.)

Beetles found in Perrowville, Va., underneath the bark of apple-trees which show decay, and thought to be the cause, are *Hymenorus obscurus* (Say). They could not have caused the injury, but were probably drawn thither to feed upon the decaying material. None of the *Cistelidæ* are known to be obnoxious.

The Stalk-Borer. (Country Gentleman, for August 3, 1882, xlvii, p. 605, c. 3 — 15 cm.)

A caterpillar, injurious to the potato crop near Syracuse, N. Y., is *Gortyna nitela* Guen.; often destructive to corn; cannot be reached by external applications: may be removed by cutting into the stem of the potato; burning the vines for killing the pupæ remaining in them recommended.

Wire-Worms infesting Potato Vines. (Country Gentleman, for August 10, 1882, p. 625, c. 2 — 18 cm.)

Reported from Scarsdale, N. Y., as first boring into the vines near the ground, and later into the tubers ; belong to the *Elateridæ* but the species unknown ; habits of the beetles ; crops attacked by the larvæ ; no effectual remedy for them known; reference to Dr. Fitch's paper on wire-worms in his 11th Annual Report.

The Horn-Tail Borer — Tremex Columba (*Linn.*). (Country Gentleman. for August 10, 1882, xlvii, p. 625, c. 2–3 — 26 cm.)

A maple tree at Poughkeepsie, N. Y., shows fifty or more holes in the trunk bored by this insect ; how the egg is inserted ; operations of the larva ; the oak, elm and the sycamore also attacked ; *Glycobius speciosus* more injurious to maples ; *Rhyssa lunator* and *R. atrata* associated with the *Tremex* as parasitic upon it.

The Spotted Horn-Bug. (Country Gentleman, for August 17, 1882, xlvii, p. 645, c. 2 — 24 cm.)

A beetle with a very offensive odor, injuring the foliage of ash trees in Perrowville, Va., is the *Dynastes Tityus.* The odor of the specimens sent is so intolerable that they cannot be kept within doors. The beetle is described ; the larva feeds upon decaying trees. The insect is rare in Pennsylvania, and is not known to occur in New York. The depredations of the beetle upon the foliage of the various trees which it attacks, may be arrested by showering with Paris green or London purple.

Mites infesting a Poultry House. (Country Gentleman, for August 17, 1882, xlvii, p. 645, c. 2–3 — 13 cm.)

The species is not recognized — may be the same that infests the nests of sitting-hens. They may be killed with kerosene in water applied with a syringe or force-pump to every part of the hen-house. If occurring in a close building, fumigation with sulphur would be efficient.

A New Worm in Apples. (Country Gentleman, for September 21, 1882, xlvii, p. 745, c. 2–3 — 21 cm.)

A white worm in early apples reducing the interior to pulp while the outside is fair, in Ascutneyville, Vt. ; is not determinable from the brief statement sent ; may perhaps be the *Sciari mali ;* operations of this species as given by Dr. Fitch ; its larva and imago described. [Is probably *Trypeta pomonella.*]

The Black Blister Beetle — Epicauta Pennsylvanica (*DeGeer*). (Country Gentleman, for September 21, 1882, xlvii p. 745, c. 3 — 15 cm.)

The beetle injures carrots and cabbages in Baltimore, Md., which have not previously been recorded as among its food-plants ; habits of the beetle ; for destroying them beat them into a vessel with kerosene and water or dust with pyrethrum, as was successfully done at Ithaca, N. Y., to large numbers defoliating a passion-vine.

The Hag-Moth Caterpillar. (Country Gentleman, for September 21, 1882, xlvii, p. 745, c. 3–4 — 20 cm.)

The larvæ of *Phobetron pithecium* (Sm.-Abb.), feeding on a crab-apple tree, described ; its cocoon and its appendages ; the larva found also on cherry-trees ; its power of stinging ; unreasonable fear of most caterpillars.

A New Apple Insect — Amphidasys cognataria *Guen.* (Country Gentleman, for October 5, 1882, xlvii, p. 785, c. 2–3 — 34 cm.)

Injuring apple-trees in Chelsea, Wis.; description of larva ; now first noticed on apple ; other food-plants ; characteristics of the *Geometridæ;* will probably not prove very injurious ; may be removed from trees by jarring.

Destructive Elm-leaf Beetle — Galerucella xanthomelæna *Schrank.* (Country Gentleman, for October 12, 1882, xlvii, p. 805, c. 1-2 — 38 cm.)

Identified from leaf injuries in Bound Brook, N. J. Depredations in N. Y., N. J. and southward; its appearance. Spraying with Paris green and water, jarring and boxing about the base of trees recommended for its destruction.

A New Principle in Protection from Insect Attack. [Read before the Western N. Y. Horticultural Society, at its Annual Meeting, January 25, 1882.] (Proceed. Western N. Y. Horticultural Society, for 1882, pp. 52–66. Separate, with one-half title p. cover, pp. 15 [March, 1882.])

Preventives preferable to remedies ; strongly odorous substances, available for the prevention of egg-deposit; insects guided in oviposition by the sense of smell rather than by sight; acuteness of this sense in man and animals; the probable location of smelling organs in insects; instances of attraction, sexual and otherwise, by odors; scent-organs and their importance ; counterodorants and how they may prevent egg-deposit ; the results of protection by this method ; the aim of practical entomology.

A Rose Leaf Insect. (Country Gentleman, for March 1, 1883, xlviii, p. 169, c. 2 — 17 cm.)

A caterpillar feeding on rose leaves in a green-house in Westchester Co., N. Y., identified as *Penthina nimbatana* (Clem.). The caterpillar briefly described, its habits mentioned, and method proposed for its destruction.

Of Interest to Flower Growers— A New Enemy Found. (Troy Daily Times, for April 2, 1883.)

Discovery of a species of caterpillar, feeding in the green-house of Dr. R. H. Sabin, Troy, N. Y., upon heliotrope, geranium, wandering jew (*Tradescantia*), etc. ; features of the caterpillar and cocoon ; is probably *Plusia dyaus* Grote, although seemingly differing somewhat from that species.

The Bean Weevil. (Country Gentleman, for April 19, 1883, xlviii, p. 317, c. 3 — 16 cm.)

Identification of *Bruchus fabæ* Riley, from Delhi, N. Y. For history of the species, reference is made to C. G. of December 8, 1881. The apartments where they are found at this season should be searched for the living individuals, which should be killed ; and all beans should be kept during the winter in tight vessels or bags to prevent the escape and distribution of the beetle.

Thousand-Legged Worms in a Nursery — Julus cœruleocinctus *Wood*. (Country Gentleman, for May 24, 1883, xlviii, p. 421, c. 2 — 26 cm.)

Identification of the species sent from Geneva, N. Y., May 10th. How they differ from "wire-worms" with which they are often confounded. *Julida* live usually on decaying vegetables, but this species feeds also on living vegetables, often on potatoes. It will probably not injure apple-tree roots, being more of a surface feeder, coming abroad at night, when it may be trapped by means mentioned. Gas-lime recommended for its destruction. The eggs, the young and habits are referred to.

Curious Ichneumon Cocoons. (Country Gentleman, for June 14, 1883, xlviii, p. 481, c. 2–3 — 35 cm.)

Cocoons of *Apanteles congregatus* (Say) occur upon an apple-tree, in Brooklyn, N. Y. The peculiar cluster is described; the habits of the Microgasters given; and the importance of protecting, rather than destroying similar parasitic cocoons.

On an Egg-parasite of the Currant Saw-fly, *Nematus ventricosus*. (Psyche, for May and June, 1883, iv, pp. 48–51.)

First discovered by the writer in the year 1866, at Utica, N. Y. Its oviposition described. Rediscovered in 1882, at Albany, N. Y. Proves to be *Trichogramma pretiosa*, described by Riley in 1879, from eggs of the cotton-worm moth. Parasitized eggs of the Nematus have been distributed to other States and Canada for colonization. Notice of the oviposition of *Nematus ventricosus* as observed.

Rearing Lepidoptera. (Psyche, for May and June, 1883, iv, p. 53 — 13 cm.)

Notices the lepidopterological studies and especially the larval collections, and rearing from the egg, of Mr. S. L. Elliot, of New York city. His success in rearing lepidoptera has not been surpassed by any one in the United States.

Codling Moth of the Apple. (Country Gentleman, for June 28, 1883, xlviii, p. 521, c. 2 — 11 cm.)

Numerous codling-moths (*Carpocapsa pomonella*) received from the Rochester, N. Y. nurseries of Mr. Barry, with report of injuries, indicating an increase of this apple-pest in the western part of the State. The threatened increase should be earnestly combatted by the most approved methods. Paper bands around the trees and showering with Paris green in water early in the season, recommended.

An Interesting Bug. (Country Gentleman, for June 28, 1883, xlviii, p. 521, c. 2–3 — 33 cm.)

Insects sent from Burlington, N. J., for identification and habits

are larvæ (1st stage) of the *Reduviina*, which cannot be positively identified, but may be the " wheel-bug," *Prionotus cristatus* (Linn.). The
eggs received with them are described and also the larvæ. The species
deserves protection from its habit of preying on other destructive insects. Their habits are briefly given, together with an account of a
two hours' contest observed between one of the larvæ and a pupa of
Penthina nimbatana. The result of a wound inflicted by one of these
insects is stated.

Book Notice. (Psyche, for May and June, 1883, iv, p. 53 — 11 cm.)

Notices the volume on the Insects of our Fruit-Trees, by Mr. William Saunders, as soon to be published.

The Maple-Tree Scale-Insect — Lecanium innumerabilis *Rathvon*.
(Country Gentleman, for July 5, 1883, xlviii, p. 541, c. 3-4 — 66 cm.)

The scales received from Parkersburg, W. Va., June 18th, with inquiries. They belong to the *Coccidæ;* history of this species given;
the different trees attacked by it; reference to papers treating of it.
Among the remedies for the insect, are mentioned, scraping and scrub-
bing the bark, cutting down and burning badly infested trees, spraying
with a whale-oil soap solution or with a kerosene emulsion. Directions
for preparing the latter are given. The literature of the species is presented.

The Black Long-Sting — Rhyssa atrata (*Fabr.*). (Country Gentleman,
for July 12, 1883, xlviii, p. 561, c. 2-3 — 28 cm.)

Insect from Athens, N. Y., identified as *Rhyssa atrata*. Its principal
features are given. The habits of the Rhyssa of ovipositing in the
larvæ of *Tremex Columba* as usually given, has recently been questioned
by Mr. Clarkson, in the *Canadian Entomologist* — very properly so, it is
thought. Another use for the long ovipositor is suggested, based upon
an occurrence noticed by the writer.

Hairworm — Vanessa — Alaus — Gordius and Mermis. (Country Gentleman, for July 19, 1883, xlviii, p. 581, c. 1-2 — 52 cm.)

A Gordius from Bainbridge, N. Y., identified. The popular superstit-
ion of the transformation of a horse-hair into the hairworm is referred
to. The general character and modes of occurrence of *Gordius* and
Mermis are given, and the life-history, in brief, of the former. *Vanessa
antiopa*, sent as injurious to elms is characterized in its larval and butterfly stage. *Alaus oculatus* is briefly described and some of its habits
given.

An Oak Moth — Anisota senatoria (*Sm.-Abb.*). (Country Gentleman,
for July 26, 1883, xlviii, p. 601, c. 3 — 16 cm.)

Received from Roslyn, N. Y. With its name, the principal features
of the moth and the caterpillar are given. Its abundance at Karner,
N. Y., is stated, together with its several changes, and mention of con-
generic species, less injurious.

Captures of Feniseca Tarquinius (*Fabr.*). (Psyche, iv, for July–August, 1883, p. 75 — 13 cm.)

A number of examples of the species collected at Keene Valley, N. Y., about alders. Had also been taken in one example at Center, N.Y., by Mr. O. von Meske, but never before by the writer.

[Platygaster larva destroying galls of Cecidomyia salicis-batatus.] (Psyche, iv, for July–August, 1883, p. 79 — 7 cm.)

Reference to Prof. D. S. Kellicott's observations on the above, as published in the Bulletin of the Buffalo Naturalists' Field Club, for March, 1883.

[Collecting Cut-worms at evening with a light.] (Psyche, iv, for July–August, 1883, p. 80 — 10 cm.)

Notice of collections made by Mrs. Mary Treat, about dusk in the evening, from flower buds of phlox, and from beneath the ground around the roots and from the branches of a plum-tree.

A Grape Pest — Procris Americana. (Country Gentleman, for August 2, 1883, xlviii, p. 621, c. 2–3 — 18 cm.)

The insect steadily increasing in Champaign county, Ohio. The caterpillar and moth are described, and the transformations given. The best remedy for it is to look for the skeletonized leaves at the commencement of the attack, to pinch them off and crush the larvæ.

Potter-wasp Cells on Grape Leaves— Eumenes fraternus *Say*. (Country Gentleman, for August 9, 1883, xlviii, p. 641, c. 4 — 22 cm.)

Sent from Sandy Hill, N. Y., for determination. The external appearance of the cells described, and also the interior, and the contained pupa; is probably the above-named species, which is common in Eastern United States. Remarks upon the structure of the cells; they are crowded with caterpillars for the larval food; the larva matures in about a month. Reference to figures in Saunder's "Insects Injurious to Fruits," page 70.

The Frenching of Corn. (Country Gentleman, for August 16, 1883, xlviii, p. 661, c. 4 — 28 cm.)

"Frenching" results from various causes — from disease as well as insect attack. Young stalks of "frenched" corn received from Rock Hall, Md., show numerous perforations made undoubtedly by a *Sphenophorus*, and probably by *S. sculptilis*. The appearance, habits, distribution, and life-history, in part, of the beetle, is given, and the best remedies for it.

The Striped Squash Beetle. (Country Gentleman, for August 23, 1883, xlviii, p. 681, c. 2 — 15 cm.)

Gives the appearance and habits of the larva of *Diabrotica vittata*

(Fabr.), and recommends the "application of carbolic acid and water (1 part to 100) about the stems to prevent attack. If the larvæ are eating into the stem, sand saturated with kerosene oil should be spread around it, — the oil to be gradually carried into the soil to kill the larvæ.

The Carpet Bug, Anthrenus scrophulariæ. (Country Gentleman, for August 23, 1883, xlviii, p. 681, c. 2–3 — 16 cm.)

The larva reported as injuring linen and silk goods, but its feeding upon these articles has not been established, and is doubted. Benzine is named as the best remedy for its attack, and coal-tar roofing-paper, and "carbolized paper " as preventives.

The Pine Emperor Moth. (Country Gentleman, for September 27, 1883, xlviii, p. 781, c. 2–3 — 24 cm.)

The larva of *Eacles imperialis* (Drury), is sent for identification from New Jersey, where it was found feeding upon plum. It had not been hitherto recorded upon this food-plant. Its food-plants are mentioned, the caterpillar described, its habits, transformations, and the principal features of the moth given. Reference is made to further information of the species.

Saw-Fly Larvæ on Quince. (Country Gentleman, for October 4, 1883, xlviii, p. 801, c. 2 — 12 cm.)

Quince leaves in Erie, Pa., are badly eaten by the larva of a saw-fly feeding on the upper surface of the leaf. The species is not identified. A description of it is given. The larva of *Vanessa antiopa*, feeding on elm, and *Adalia bipunctata*, found on quince, are identified.

Bark Louse on Willow. (Country Gentleman, for October 4, 1883, xlviii, p. 801, c. 2–3 — 22 cm.)

The apple-tree bark-louse is found incrusting a stem of Kilmarnock willow, from Ansonia, Conn. Other plants upon which it is known to occur in addition to apple, are named. The species belongs to *Mytilaspis*, and according to Prof. Comstock, is identical with *pomorum* of Bouché. As remedies, scouring with soap-suds and a stiff brush, and showering with a kerosene emulsion are recommended.

The Chinch-bug in Northern New York. (Albany Argus, for October 10, 1883, p. 3, c. 2–3 — 90 cm. Watertown Daily Times, for October 12, 1883 [the same article copied]. Country Gentleman, for October 18, 1883, p. 841, c 2–4. [the same copied nearly entire].)

Following a letter from Mr. M. H. Smith, of Redwood, Jefferson Co., N. Y., giving an account of the discovery of the insect and the injuries committed, the report of the writer is given, after his visit to the infested locality, embracing the following heads : The insect identified — appearance of the insect — observations upon the attack — just cause for alarm — persistence of the attack — importance of arresting the attack — remedial measures recommended.

A New Enemy to the Farm.   (Albany Argus, for October 10, 1883, p. 4, c. 3 — 30 cm.)

A recapitulation of the preceding paper, slightly altered from MS. to serve as an editorial.

The Chinch-bug in New York.   (Science, for October 19, 1883, ii, p. 540 — 16 cm.)

Its detection in large numbers in St. Lawrence Co., N. Y., where it is proving destructive to timothy-grass. Its rapid increase noticed, notwithstanding that this and the past year have been unusually wet ones in Northern N. York. Its threatened spread is occasioning great alarm, but it is hoped that it may be arrested by the general use of kerosene oil emulsified.

Directions for Arresting the Chinch-bug Invasion of Northern New York.   (Circular No. 1 — October, 1883.   New York State Museum of Natural History : Department of Entomology, 8vo., 3 pp., fig.)

Narrates the features of the attack in St. Lawrence Co., N.Y., refers to the importance of arresting its extension, and recommends for the purpose: 1. Examination for detecting the commencement of the attack : 2. Burning, as directed ; 3. Plowing as directed ; 4. Harrowing and rolling ; 5. When plowing is not practicable, use gas-lime. The above to be done at once.   Further directions for attacking the spring brood are promised hereafter.

The Mole-Cricket.   (Country Gentleman, for October 25, 1883, xlviii, p. 861, c. 2 — 20 cm.)

The insect sent as being quite plentiful at Woodbury, N. J., is identified as *Gryllotalpa borealis* Burm., and described.   Its method of excavating its galleries are narrated and the injuries committed by it upon grass and garden vegetables.   Hot water poured into the burrows of the insects will kill them when they become too abundant.   In Europe they are sometimes trapped in manure pits.   The ability of the insect to swim and dive readily in the water, as communicated, had not been previously recorded.

The Ant Lion.   (Country Gentleman, for November 1, 1883, xlviii, p. 881, c. 1-2 — 33 cm.)

Larva of *Myrmeleon* sp.? received from Falls Church, Va.   Appearance of the ant-lion, structure, the pitfall constructed by it, and its operations with its prey.   Means by which the larvæ may be collected, with remarks upon the winged insect.

New Corn Pest — Megilla maculata.   (Country Gentleman, for November 22, 1883, xlviii, p. 941, c. 1, 2 — 48 cm.)

The *Coccinellidæ*, long regarded as wholly carnivorous, have lately
30

been found to be partly herbivorous. Prof. Forbes has shown that more than one-half the food of some species is vegetable. In 1874, *M. maculata* was reported as injuring corn in the milk. This habit was confirmed by Mr. Pergande in 1882. The beetles now sent from Fairfield, Conn., by Mr. Sturges, taken by him from within the kernels of corn, are the same species. The nature of the injury to corn is described, also the insect itself, and a cause for its attack suggested.

The Apple-Maggot — Trypeta pomonella. (Bulletin No. LXXV. N. Y. Agricultural Experiment Station. December 29, 1883 — 110 cm.)

An attack upon apples in Brandon, Vt., by which they are completely "honey-combed," is stated to be probably by *Trypeta pomonella*. The habits of the larva, its description, that of the fly, life-history of the species, its ravages and distribution, and remedial measures against it are given. Information (points stated) still needed of these attacks. *Sciari mali* should not be confounded with the species.

(C.)

The following valuable paper of Dr. Fitch upon some rare insects of the State of New York, was almost lost in publication, from the small number of copies published of the volume in which it appeared, and the difficulty of obtaining access to it. A few copies of the paper were also issued as separates with, with pagination of 1–11, but they are to be found to-day in hardly any of our public libraries, or even in private hands. Its republication, therefore, at this time, cannot fail of being acceptable to entomologists.

[*From the American Quarterly Journal of Agriculture and Science, May,* 1847, *vol. v, pp.* 274-284.]

WINTER INSECTS OF EASTERN NEW YORK.

BY ASA FITCH, M. D.

It is the object of the following paper, to describe those insects of Eastern New York, which occur in their perfect state in the winter, and are peculiar to that season and the early part of spring. They are objects of curiosity, as coming forth to our view in full maturity and vigor, at that time in the year when almost every other member of the animal and vegetable kingdoms is reposing in torpidity under the chilling influence of solstitial cold. In an economical aspect, they possess but little importance, their period of life being limited to that season when the field furnishes no herbage, the garden no flowers, and the orchard no fruits, on which they can prey. They are chiefly interesting, therefore, merely as objects of scientific research — as forming integral parts of that vast array of animated beings, with which the Father of Life has populated our world, and rendered it vocal with his praise.

Hence it is to the scientific rather than the agricultural reader, that the following pages are addressed. To him they will be sufficiently intelligible, without such illustrations as have accompanied our previous contributions to this Journal.

A few words respecting the analogies of the two first species here described, may not be devoid of interest to the general reader. A small insect, destitute of wings, and bearing some resemblance to a flea in its general aspect, is found in the winter season, upon the snow in the northern part of Europe, and also occurs upon the Alps and the Hartz mountains. It has been known for nearly a century, and from its singularly anomalous characters, naturalists have been much perplexed to determine in which particular family of the insect tribes it might with the most propriety be placed. Linnæus was the first to classify and name it. He regarded it as possessing more analogies with the species associated in his genus *Pa-norpa*, [*] than with any other insects, and accordingly arranged it with them, bestowing upon it the specific name *hyemalis*. But, inasmuch as it differed from the Panorpidæ in some prominent particulars, such as possessing the faculty of leaping, and being furnished with an ovipositor similar to many grasshoppers and crickets, Panzer, at a subsequent day, placed it under the genus *Gryllus*. More recent naturalists, however, have concurred in the propriety of the location originally given by Linnæus, and to obviate, in some degree, the incongruity of its situation, Latreille was induced to construct for it an independent genus, placed beside Panorpa, to which genus he gave the name *Boreus*. The *hyemalis* has remained to this day the sole species of this genus, no other insect having similar characters, having been discovered in any part of the world. Two years since, in the month of March, searching carefully upon the melting snow, to find if possible in this vicinity, a rare and singular insect which has been lately discovered in Canada — the *Chionea valga*, a fly destitute of wings — though unsuccessful, my labors were rewarded with an equally acceptable return, an insect cogeneric with the curious *Boreus hyemalis* of Europe. Since that time, I have met with numerous specimens, and have also found in the same situations, several individuals of a third species pertaining to the same genus. From these specimens I draw the following detailed characters of the

Genus BOREUS, *Latreille.*

Polished and shining. *Head* sunk into the thorax to the eyes, which are prominent; ocelli wanting. *Rostrum* long-conical, twice or thrice as long as the head from which it gradually tapers, projecting downwards at right angles with the body, or more or less inclined backwards under the breast, its front side clothed with minute hairs. Maxillary *palpi* reaching beyond the tip of the beak; terminal joint longest and slightly thicker than the others, long ovate: basal joints cylindrical, half as long as they are broad. *Antennæ* inserted in the middle of the front, their bases nearer to the margin of the eyes than to each other, reaching half the length of the abdomen in the females and to its tip in the males, thickly set with very short minute hairs; filiform, hardly thicker toward their tips, composed of twenty-three joints; two basal joints thickest, the first subcylindric, the second obovate; succeeding joints short-cylindric, compact; terminal joint ovate. *Thorax* cylindrical, scarcely as broad as the head. *Wings, in the males,* rudimentary and not adapted for flying. Upper pair represented by two coriaceous pseud-elytral scales which reach rather more than half the length of the abdomen; these are broadest at their base and gradually taper to an acute point, the length being over four times as great as the breadth; they are very convex above and concave on their under sides, and thus when detached, bear some resemblance to the chaff-scale or glume of a small kernel of grain ; the apex is armed with a straight thorn-like spine which is directed backwards and downwards; the inner margin is studded with a row of small teeth, which are longer and more distinct toward the apex of the pseud-elytron; these teeth are inclined backwards, and at their points they are strongly curved in the same direction; both the outer and inner margins are minutely ciliated with short hairs. The under wings are represented on each side by a curved bristle which lies under the pseud-elytron and within its concavity; it scarcely exceeds the pseud-elytron in length, is slightly dilated at its base, curves inwards and downwards, is almost hooked at its tip, and gives off an occasional short hair. *In the female* the wings are entirely wanting, the only vestiges of them being two minute scales occupying the place of the upper pair; these scales are circular and scarcely the hundredth part of an inch in diameter in *B. nivoriundus*, slightly elongated and a third smaller in *B. brumalis ;* they are convex above and concave beneath, and attached to the thorax by a short broad pedicel; their edges are ciliated with minute hairs : their upper surface is also thickly set with very short, erect hairs, and is crossed by an elevated rib or slight keel. *Legs* long, particularly the posterior pair, the length of which exceeds that of the body; their several joints cylindric and densely clothed with short minute hairs; the first tarsal joint half as long as the tibia, the four remaining joints successively shorter, terminated by two small, slender, simple hooks. *Abdomen* oval, depressed when exsiccated, the segments distinctly marked by strongly impressed transverse lines, and clothed with fine appressed hairs; in the males it is nearly cylindrical, but little broader than the head, truncated as it were at its apex and turned upwards; tip of the last segment furnished with two stout sharp-pointed hooks, each with an acute tooth in the middle of its inner edge, and pilose along its outer edge; these hooks are susceptible of being extended in a line with the body, but are commonly strongly recurved upon the back shutting down upon and grasping a small scutel-like process which projects upwards at the base of this segment. They are thus recurved in coition, the male organ being exserted from between their bases. *Ovipositor* robust, about half as long as the abdomen of the female, projecting backwards in a line with the body, composed of a three-jointed semicylindrical piece above, and two ligulate valves below; the latter have their lower edges held in contact, thus forming a little gutter, and on the under-side toward their tips they are finely serrated; of the upper piece, the middle joint is much the longest, and is lined beneath on its concave side with a membrane which becomes distended with fluid when the abdomen is pressed upon; the short terminal joint is susceptible of being inclined obliquely downwards, thus, at least partially, closing the end of the ovipositor; the upper and lower pieces are widely separated in coition to enable the tip of the male abdomen to approximate that of the female.

1. BOREUS NIVORIUNDUS. *The Snow-born Boreus.*

Shining black or brownish-black ; rudimentary wings, thorax above, with the rostrum and ovipositor excepting their tips, fulvous; legs dull fulvous.

Length, male twelve-hundredths of an inch; female, o.15, or including the ovipositor, o.18.   .

Head black, highly polished, glabrous. Eyes black. Rostrum fulvous and feebly diaphanous, the mouth and palpi black. Antennæ black, two basal joints sometimes fulvous-brown. *Thorax* black on the sides, above varying in color from dull fulvous to cinnamon-yellow, the basal half of the prothorax being black. *Abdomen* black, brownish black, or dull ful-vous-brown; terminal segment fulvous or cinnamon-yellow, its hooks in the males cinnamon-yellow, their tips and teeth black and highly polished; ovipositor in the females diaphanous, fulvous, sometimes inclining to rufous, black at its tip. Rudimentary *wings* cinnamon-yellow; in the males often of a duller hue toward their tips; rudimentary inferior wings in the males of the same color as the superior. *Legs* lurid-yellow and sub-diaphanous, with a slender black annulus at each of their articula-tions; three last joints of the tarsi wholly black.

Closely allied to the *B. hyemalis*, which, however, appears from Ram-bur's Neuroptera, the Penny Cyclopædia, and the beautiful colored figure in Westwood's Introduction, the only definite authorities to which I am able to refer, to have the basal two-thirds of the antennæ of a russet color, and the rudimentary wings and the legs strongly inclining to red. Our species presents no tinge of rufous, except sometimes in the ovi-positor; and the antennæ, black to their bases, is a decided distinctive mark.

This insect is by no means rare, being found upon the snow in forests in warm days, so early as December, and becoming more common as the season advances. I have met with it the most plentiful in April, when there has been a fall of snow in the night, succeeded by a warm forenoon of bright sunshine. Appearing so suddenly, in numbers, upon the clean, dazzling white surface thus spread over the earth, at the first thought it seems to be literally bred from the snow. I have not yet searched for it in the moss of tree-trunks, but doubt not that like the European insect, ours will also occur in this situation. When observed upon the snow, it is almost always stationary; and when approached by the hand, it com-monly makes a leap, to the distance of a few inches only, its saltatory powers appearing but feeble.

2. BOREUS BRUMALIS.   *The Mid-winter Boreus.*

Polished deep black-green; legs, antennæ, rostrum, and ovipositor black; rudimentary wings brownish-black.

Length, male o. 10; female o.12, or including the ovipositor o.15.

This species presents no very obvious characters beyond those already given. Its body is highly polished, shining even with a metallic lustre whilst the eyes, antennæ, rostrum, and legs, reflect the light but feebly. The ovipositor is pure black, but equally splendent with the black-green abdomen. The scales which occupy the place of the wings in the females are but faintly perceptible, appearing like two minute greyish-black spots on the thorax. In the living insect, there is a light fulvous vitta, obvious to the naked eye, along each side of the abdomen, at the lateral suture; this is frequently obliterated or but imperfectly discernible in the dried specimen.

So far as I have at present observed, this appears abroad earlier in the season, and in colder weather than the preceding, though occasionally found associated with it on the last snows that fall in the spring. It is much less common than the other.

3. PERLA NIVICOLA. [*] *The Small " Snow-fly."*

Black ; wings grey, unclouded, a third shorter than the abdomen in the males, a third longer in the females.

Length 0.20, wings expand 0.45; males smaller.

Head shining, clothed with very short, fine hairs. Palpi brownish-black, sub-diaphanous. Antennæ reaching half the length of the wings, black, setaceous, about thirty-jointed ; joints obconic, basal one largest. *Pro-thorax* flattened, its margins more smooth and shining, its disk rugulose, with a few shallow impressions ; an impressed transverse line near the base and another near the apex. *Abdomen* shining, with a broad pale fulvous dorsal vitta which does not extend on to the two last segments ; ven-ter with a tint of obscure pallid at base. Setæ as long as the abdomen, black, setaceous, clothed with short whitish hairs ; joints from thirteen to about eighteen in number, obconic, gradually shorter toward the base. *Legs* black, joints cylindric. Tibiæ obscure pale brown except at the tips, subdiaphanous, grooved longitudinally. Tarsi, basal joint longest, second joint very short. *Wings* reaching half the length of the setæ, finely cili-ated at their tips and along their inner margins ; gray, diaphanous, im-maculate; nervures black, robust, and very strongly marked, particularly on the upper pair which have five closed cells in the disk. The male is smaller, with the wings reaching but two-thirds the length of the abdo-men, its palpi and entire tergum black, and the tibiæ darker than in the female.

On warm days in the latter half of winter this species may be observed crawling with hurried steps upon the snow. It becomes most numerous about the time the snow finally disappears, and is then often seen on shrubs, fences, and buildings, and not unfrequently finds its way into our houses. It is extremely common, occurring most abundantly in the vi-cinity of streams of water, in which element the previous stages of its existence are passed. When first excluded from its pupa state, it is of a pale yellowish color, but gradually changes to black, this change com-mencing upon the thorax. Copulation occurs immediately after the female comes from the pupa state.

[* Is *Capnia pygmæa* (Burm.) Pictet: Hist. Nat. Ins. Neurop., 1841, p. 324, pl. 40, figs. 1–3.]

4. Nemoura nivalis. *The Large "Snow-fly." The "Shad-fly."*

Black; wings griseous, faintly banded, double the length of the abdomen.

Length, males somewhat under, females over half an inch; wings expand about an inch.

Head covered with minute whitish hairs, which are longer and more obvious beneath the bases of the antennæ and around the mouth. Vertex with an obtusely impressed transverse line immediately back of the two posterior stemmata, and a longitudinal medial one, reaching from the former to the neck. Antennæ black, clothed with very short minute hairs, slender, setaceous, as long as to the tips of the wings in the males and somewhat shorter in the females, composed of about sixty joints; basal joint short-cylindrical, its diameter double that of the third and following joints; second joint intermediate between the first and third in diameter, its length and breadth about equal; the remaining joints obconic, gradually diminishing in diameter and increasing in length toward the tips. Palpi clothed with very short, minute hairs, black; basal joints of the maxillaries lurid and slightly diaphanous, penultimate joint rather the shortest and obconic, the joint preceding it longest and obconic, the terminal joint oval, and scarcely as thick as the others. *Prothorax* square, in the females scarcely broader than it is long, somewhat narrower anteriorly, posterior angles rounded, all the margins slightly and obtusely elevated, the posterior one more obviously so, often with a dull fulvous spot at the base, or with this color spread over the posterior part of the raised margin, and more rarely a similar spot at the middle of the apex; disk sometimes showing an impressed transverse line, and a longitudinal dorsal stria. Exposed portion of the *mesothorax* much elevated above the plane of the prothorax, forming a transverse ridge between the bases of the wings; clothed with short hairs; often with traces of dull fulvous around the wing-sockets; the portion of the mesothorax and metathorax covered by the wings smooth and shining. *Abdomen* reaching but half the length of the wings; sutures of the tergum in the female more or less widely marked with dull rufous; tip, in the female only, furnished with two short, filiform setæ, scarcely equalling in length the segment to which they are attached; setæ pale lurid, sub-diaphanous, hairy, composed of about eight joints. Each segment of the venter with two transverse impressions, one situated toward each posterior angle. Male organ exserted, forming a conical lurid point near the base of the last ventral segment. *Femurs* cylindrical, black, clothed with white hairs, which are longer and more distinct in the females, inner side with a narrow deep groove which is dilated toward the apex. *Tibiæ* cylindrical, about half the diameter of the femurs, grooved, lurid-brown, diaphanous, the ends and inner sides black; apex slightly incurved and armed with two short spines on the inside. *Tarsi* black, composed of three joints, whereof the middle one is slightly shorter; two claws and an intervening pellet at the tips. *Wings* griseous, when closed showing faintly two paler bands, one near the middle and the other back of it; edges ciliated with fine, short hairs. Upon wings diaphanous, gray, faintly marked with a darker cloud back of the middle, and another occupying the tips, but not reaching to the edge, these clouds becoming wholly obliterated in cabinet specimens; nervures black. Lower wings gray, sub-hyaline, nervures black.

When recently excluded from the pupa, the abdomen, except at its tip, is of a dull rufous color; this gradually becomes darker, and finally pure black. For a time after the venter has become wholly black the tergum continues dull rufous with a black band on each segment, which band does not reach the lateral margins. These bands increase in size, and at length the whole tergum is overspread with pure black.

It is not uncommon to meet with specimens of this and the preceding species, infested with a minute parasite of the family *Acaridæ*. These parasites are of a bright vermillion-red color, and fix themselves, one or more, at the sutures of the tergum, not quitting their hold after the death of the insect, unless disturbed.

This species begins to appear, soon after the small Snow-fly is first met with. It occurs in the same situations, is nearly as abundant, and re-mains for a time after that has disappeared. One of the purposes served by these prolific insects in the economy of nature, doubtless is, to supply with food the fish of our streams, at this early period of the year. The larger of these species continuing to be abundant when the shad first come into our rivers, has evidently received one of its popular designa-tions in allusion to this fact.

We regard this as the American analogue of the European *Nemoura nebulosa*, Linn. But from several points in the extended description of that species given by M. Ramber (*Suites à Buffon, Insectes Nevroptères*, Paris, 1842), it is quite obvious that ours is a distinct species.

5. CULEX HYEMALIS. [*] *The Winter "Musketoe."*

Thorax cinereous, with a broad black vitta on each side; extreme tips of the wings and two spots on their anterior margins black, with two in-tervening sericeous yellowish white spots.

Length 0.22; to the tips of the wings 0.28, or including the beak 0.39.

Head cinereous-pubescent, occiput black-pubescent. Proboscis black, its apex cinereous. Palpi black, the tips varied with gray. Antennæ black, tips brown. *Thorax* cinereous-pubescent, with a broad rufous-black vitta on each side, passing above the wing-sockets; the vitta often edged on its upper side with yellowish-white; a very slender, black, dorsal line, often partially obsolete. Scutel glabrous, dark brown. Poisers black, their pedicels white. *Abdomen* clothed with longish gray hairs, black or dark brown, with two rows of whitish spots on each side; in the males obscure white, the posterior margins of the segments black. *Wings* sub-hyaline, with two blackish spots on the anterior margin, separated by a

---

[* Is *Anophales quadrimaculatus* Say: in Long's Exp., Append., ii, 1824; Compl. Writ., i, 1859, p. 241.]

conspicuous glossy yellowish-white spot ; inner spot with a strong notch on its posterior side which is formed by a yellowish-white dot, and a similar dot is placed on the inner side of this spot; outer spot with an oblique yellowish-white band on its outer side, beyond which at the tip of the wing, is a slight blackish transverse spot. Under a magnifier, these spots are found to be produced by the colors of the scales upon the nerves of the wings, which scales are regularly and beautifully dyed with black and yellowish-white, as follows: the posterior or anal nerve has black scales the last half of its entire length, and also at its base ; the next or interno-medial nerve, which forks in its middle, is clothed through-out with black scales, including both its branches; the next or externo-medial has black scales on the basal fourth of its length, two broad annuli of black scales on its middle, another annulus at its fork, and a fifth ser-ies at the tips of each of its branches; the next is clothed with black scales through its entire length ; the next is black where it first becomes plainly visible in the middle of the wing, again for a short distance after the origin of the preceding nerve, again for a considerable space of its fork, and again at the apex of its posterior branch only ; the costal and the marginal nerves have black scales from their bases; these become much more dense at the black spots of the anterior margin, and are replaced by yel-lowish scales only between these spots and beyond the entire one. *Legs* black ; femurs pale toward their bases; tips of femurs and of tibiæ whitish. Coxæ pale.

The Winter Musketoe is met with in the last days of autumn and again for a short time in the first days of spring, and specimens are occasionally found in any of the winter months. It is a somewhat rare insect, which no one can fail to distinguish clearly by the marks on its wings as above described.

6. CHIRONOMUS NIVORIUNDUS.   *The Snow-born Midge.*

Black ; poisers obscure-brown ; wings pellucid-cinereous, their anterior nervures blackish.

Length about 0.15 to the tip of the abdomen in the males ; females a third shorter.

This species is black throughout, and clothed with fine black hairs. The *thorax* has three slightly elevated longitudinal ridges immediately forward of the scutel. The *wings*, when the insect is at rest, are held against the sides of the abdomen, often vertically in the males, but more commonly in the females with their inner margins in contact, thus form-ing a steep roof covering the back. They are diaphanous, of a cinereous tinge, and feebly iridescent. Their inner margins toward their bases are slightly arcuated. The submarginal or postcostal nervures, those which bound the closed basillary cell, and which proceed from this cell to the margin, are particularly obvious, being of a blackish color, excepting the nerve which proceeds from the inner angle of this cell to the apex of the wing, which, with the nervures inside of it, scarcely differ in color from the surface which they ramify. The *poisers* are obscure-brownish, truncated at their apices, the capitulum being in the form of a reversed triangle. The *abdomen* in the females is shorter than the wings, somewhat compressed, approaching to an ovate form when viewed laterally, with the venter often of a dull brownish tinge ; in the males it projects beyond the tips of the wings, is slender, cylindrical or very slightly tapered toward the tip, with some of the terminal segments separated by a strong contraction.

This is a very common species, appearing upon the snow in the winter season, and upon fences, windows, etc., in the fore part of spring, the males and the females being about equally numerous. The beautiful plumose antennæ of the former distinguish them at a glance from all other insects abroad at this season. At times they may be met with in immense swarms. April 27th, 1846, in a forest, for the distance of a fourth of a mile, they occurred in such countless myriads as to prove no small annoyance to the passer, getting into his mouth, nostrils and ears at every step, and literally covering his clothing. These had probably hatched from the marshy border of an adjoining lake, on this and the preceding days, the weather having been remarkably warm and dry. The wings appear to be more hyaline and iridescent in those individuals that come forth earliest, but I am unable to detect any marks by which they may be characterized as specifically distinct from those which appear at a later day.

7. TRICHOCERA BRUMALIS. *The Mid-winter Trichocera.*

Brownish-black ; wings and legs pallid at their bases ; poisers blackish, their pedicels whitish.

Length of the male 0.18, of the female 0.25, the wings expanding twice these measurements.

Thorax with an obscure grayish reflection. *Abdomen* in the males cylindrical, slightly narrower toward the tip, in the females elongated-oval and pointed at the tip ; each segment with a strongly impressed transverse line in its middle, and the posterior margin elevated into a slight ridge. *Ovipositor* fulvous, sometimes tinged with blackish. *Wings* hyaline, faintly tinged with dusky ; inner margins ciliated with quite short hairs; nervures blackish. *Legs* very long, slender, and fragile, blackish ; femurs brown, gradually paler toward their bases.

Common in forests in the winter season, coming out in warm days, flying in the sunshine, and alighting upon the snow, its wings reposing horizontally upon its back when at rest. Even when the temperature is below the freezing point, and the cold so severe as to confine every other insect within its coverts, this may be met with abroad upon the wing. It is a plain, unadorned species, closely allied in its character and habits to the European *T. hyemalis,* but in a number of impaled specimens before me, I can detect no stripes or bands upon the thorax ; whilst the very obvious character of the legs and wings being pallid at their bases, I do not find mentioned as pertaining to that species.

8. PODURA NIVICOLA. *" The Snow-flea."*

Black or blue-black ; legs and tail dull brown.
Length o.o8.

Body black, covered with a glaucous blue-black powder but slightly adherent, and sparingly clothed with minute hairs ; form cylindrical, somewhat broader toward the tail. *Antennæ* short and thick, longer than the
head. *Legs* above blackish, beneath dull brown and much paler than the
body. *Tail* of the same color with the venter, shortish, glabrous on its
inner or anterior surface, with minute hairs on the opposite side ; its fork
brownish.

Though found in the same situations as the European *P. nivalis*, ours is
a much darker colored species. Say's *P. bicolor* is a larger insect than the
one under consideration, and differs also in size and in the color of the tail
or spring. From the habits of the present species, we should infer that it
might be abundant in all the snow-clad regions of the northern parts of
this continent ; it may, therefore, prove to be identical with the *P. humicola*
of Otho Fabricius (Fauna Groenlandica), of which we are unable to refer
to any but short and unsatisfactory descriptions, which do not coincide
well with our insect.

This is an abundant species in our forests in the winter and fore part of
spring. At any time in the winter, whenever a few days of mild weather
occur, the surface of the snow, often, over whole acres of woodland, may
be found sprinkled more or less thickly with these minute fleas, looking
at first sight, as though gunpowder had been there scattered. Hollows
and holes in the snow, out of which the insects are unable to throw themselves readily, are often black with the multitudes which here become imprisoned. The fine meal-like powder with which their bodies are coated,
enables them to float buoyantly upon the surface of water, without
becoming wet. When the snow is melting so as to produce small rivulets
coursing along the tracks of the lumberman's sleigh, these snow-fleas
are often observed, floating passively in its current, in such numbers as
to form continuous strings ; whilst the eddies and still pools gather them
in such myriads as to wholly hide the element beneath them.

GENERAL INDEX.

A.

Aaron [S. F.], referred to, 203.
Abbotii, Thyreus, 214.
abieticolens, Adelges, 185.
    Chermes, 185.
abietis, Adelges, 185.
    Chermes, 185.
Acanthia lectularia, 17, 152, 154.
Acari, 36.
Acaridæ, 241.
Achorutes boletivorus, 206.
    marmoratus, 206.
    nivicola (see Snow-flea), 203-206, 244.
    pratorum, 206.
    purpurascens, 207.
    Texensis, 206.
Acrididæ, a family of the Orthoptera, 187.
Acridium hemipterum, 187.
    marginatum, 187.
    Virginianum, 187.
    viridifasciatum, 187.
Acronycta Americana, 77.
Actias Luna, 39.
Adalia bipunctata, 232.
Adelges, abieticoleus, 185.
    abietis, 185.
    coccineus, 184.
    strobilobius, 184.
Ægeria cucurbitæ, 57.
    exitiosa, 6, 60, 216.
    tipuliformis, 60, 216.
Ægeriadæ, family of, 60.
    pupal armature of, 216.
æqualis, Stenobothrus, 196.
æstiva, Dendrœca, 8.
Agelæus Phœniceus, 189.
Alalantæ, Microgaster, 39.
Alaus oculatus, 230.
Albany *Argus* cited, 43, 149, 158, 232, 233.
Albany *Evening Journal* cited, 69, 85, 189, 190, 223.
Aletia argillacea, 220.
albipennis, Bibio, 110.
Alder blight, 181.
Aleyrodes, Trichogramma bred from, 220.
Allorhina nitida, 114.
Alucita cerealella, 102, 107.
    granella, 107.
ambulans, Lipura, 208.
Americana, Acronycta, 77.
    Clisiocampa, 83.
    Procris, 231.

American Agriculturist referred to, 102, 149.
    Association for the Advancement of Science, referred to, 41, 64, 213, 217.
    Entomologist, reference to, 46, 52, 89, 93, 111, 117, 122, 125, 126, 127, 129, 132, 142, 149, 168, 203.
    Journal of Horticulture, referred to, 117, 118.
    Journal of Science and Arts, cited, 57, 59, 167.
    Naturalist, reference to, 3, 83, 124, 131, 141, 208.
    Quarterly Journal Agriculture and Science, 203, 235.
Americanus, Coccygus, 82.
ammerlandia, Scutelligera, 116.
Ampelis cedrorum, the cedar-bird, 7.
Ampelophila sp., 23.
Amphicerus bicaudatus (see Apple-twig borer), 125-132.
Amphidasys cognataria (see Currant Amphidasys), 97-101, 227.
Anacampsis cerealella, 102.
Anasa tristis, 29, 165.
anchorago, Stiretrus, 146.
Androconia of Butterflies, 214.
Angoumois moth, 102-110.
    at the State farm at Geneva, 102.
    bibliography, 102.
    distribution, 109.
    food-plants, 106.
    history in Europe and America, 104.
    life-history, 106-108.
    mite parasite on, 110.
    moth described and figured, 105.
    natural enemies, 110.
    operations in corn, 103, 104.
    parasites on, 110.
    remedies, 109.
    synonymy, 102.
Angus, James, on the 17-year cicada, 171.
    Psylla buxi from, 18.
angustata, Nysius, 166.
Anisopteryx vernata, 7, 27, 44, 224, 225.
Anisota senatoria, 230.
Annales de l' Agriculture Francaise, 110.
Anobium tesselatum, 202.
Anophales quadrimalculatus, 241.
Anthomyia brassicæ, 28.
    raphani, 28.
Anthomyiidæ, 35, 46, 225.
Anthrenus scrophulariæ, 46, 47, 138, 226, 232.
    varius, 138.

Fly-weevil (Angoumois moth), 105.
Forbes, Prof. [S. A.], cited, 8, 40, 112, 113, 157, 234.
 Reports on the Insects of Illinois, 69, 102, 110, 149, 154, 156.
Forest tent-caterpillar, 83.
Formica rufa, 117, 186.
Foster, Il. J., weevil-marked plums from, 13.
Frames with netting cover for protecting squash-vines, 64.
fraterna, Tetrastichus, 79.
 Trichogramma, 79.
fraternus, Camaronotus, 186.
 Eumenes, 231.
Frenching of corn, 15, 231.
French [Prof. G. H.], references to, 57, 69, 102.
Fruit-Growers' Journal, cited, 121.
fugitiva, Limneria, 41.
Fuller, A. S., on harlequin cabbage-bug, 56.
 referred to, 142.
Fulleri, Aramigus, 142-144.
Fuller's rose beetle, 142-144.
 beetle described and figured, 143.
 bibliography, 142.
 discovery, 142.
 distribution, 143.
 early stages figured, 143.
 food-plants, 144.
 green-house pest, 142.
 remedies, 144.
 transformations, 143.
Fumigation with charcoal gas for Angoumois moth, 109.
 with sulphur recommended. 18, 227.
funebris, Eurytoma, 2.
Fungus-feeding fly, 13.
Fungus on house-fly, 179.
 on peach twigs, 7.
 quinces, 11, 13.
 seventeen-year Cicada, 179.
 silk-worm, 179.
fur, Ptinus, 138.
fusca, Lachnosterna, 3, 41, 224.

G.

galbula, Icterus, 82.
Galerucella xanthomelæna, 228.
gallæsolidaginis, Gelechia, 39.
Gall insects, 31.
Gardener's Monthly and Horticulturist, 19.
Gas-lime for killing insects, 21, 63, 162, 163, 229, 233.
 to prevent insect attack, 162.
Gasoline for insect attack, 200.
Gas-water, for root-insects, 35.
Gelechia cercalella, 102.
 gallæsolidaginis, 39.
gelechiæ, Apanteles, 39.
 Microgaster, 39.
 Pteromalus, 110.
Geometers, 91.
Geometridæ, 91, 95, 97, 228.
 habits of the larvæ, 98, 101.
Gillette, C. M., on the green-striped locust, 188, 189.
Girdling twigs by Orgyia larva, 87, 88, 89.
Glaucopidæ, anal appendage in some, 214.
Gleason, A. H., on the green-striped locust, 195.

globosus, Aphritis, 116.
 Microdon, 116.
Glover [T.], Manuscript Notes from my Journal, 45, 111, 112, 116, 117, 145, 146, 149, 165, 166.
 in Reports of the Commission of Agriculture, 19, 102, 113, 126,129, 165, 167, 180, 188, 203.
 reference to, 106, 166.
Glycobius speciosus, 227.
Goding, Dr. F. W., on Trypeta pomonella, 121.
Goff, E. S., on the squash-vine borer, 67 68.
Gomphocerus infuscatus, 188.
 radiatus, 188.
 viridi-fasciata, 188.
Goniaphea ludoviciana, the rose-breasted grosbeak, 8.
Gooseberry fruit-worm, 10.
Gordius, a hair-worm, 230.
Gortyna immanis, 35, 41.
 nitela, 226.
Grain Aphis, 225.
 pest, 137.
granaria, Aphis, 225.
granella, Alucita, 107.
 Œcophora, 102.
granellus, Ypsolophus, 102.
granulata, Tettix, 197.
Grape bagging to prevent insect attack, 32.
 berry moth, 33.
 curculio, 33.
 Phylloxera, 21, 22.
 seed midge, 32.
Grapta and Vanessa, sexual features of, 214.
Grasshoppers, 3, 34, 188-198, 223.
Grasshoppers, midwinter, 188, 223.
Green fly, 31.
Green-striped Locust, 187-198, 223.
 appearance in Genesee county, 189.
 appearance in Westchester county, 189.
 bibliography, 187-8.
 description of the insect, 191.
 dimorphic forms, 191.
 double-brooded species, 195.
 eggs of, 193.
 habits, 192, 193, 195.
 meteorological conditions causing its appearance, 190.
 midwinter appearance, 188.
 molting operation, 193, 194.
 natural history, 192-195.
 other double-brooded species, 196.
 remedies, 197-8.
 synonymy, 187, 188, 192.
 winter appearance not alarming, 190.
grossulariæ, Pempelia, 10.
 Pristophora, 5.
Grote [A. R.], referred to, 57, 89, 92, 94, 97, 215.
Gryllotalpa borealis, 233.
Gryllus hyemalis, 236.
 Locusta chrysomelas, 187.
 Virginianus, 187.
Gypsum and kerosene for squash-bugs, 29.

H.

hæmorrhoidalis, Heliothrips, 56.
Hagen, Dr. H. A., on Tribolium ferrugineum, 138.

White-winged Bibio :
  transformations, 114.
Whitman, L. L., on apple-maggot, 121-2.
Whitney [C. P.], on Thyreus larvæ, 215.
Willard, S. D., saw-flies from, 5.
Williston, Dr. [S. W.], on Microdon bul-
  bosus, 116, 117.
Winter Insects of Eastern New York, 203,
  205, 235-244.
Winter Musketoe, 241.
Wire-worms, 35, 226.
Wollastonia quercicola, 53.
Wood-ant, 186.
Woolly-aphis of the apple-tree, 181.
  of the elm, 181.
  of the oak, 181.
Worthington, C. E., collection of Gortyna
  immanis, 42.

X.

xanthomelæna, Galerucella, 228.
Xyleborus cælatus, 54-5.
Xylotocus bivittatus, 54.

Y.

Yellow-billed cuckoo, 82, 83.
Yellow-necked apple-tree caterpillar, 83.
Yellow woolly-bear, 83.
Ypsolophus grancllus, 102.
yuccasella, Pronuba, 215.

Z.

zeæ, Sphenophorus, 52.
Zebra caterpillar, 1.
Zeuzera, 216.

PLANT INDEX.

34